Science: Entry Level Certificate

AQA Specification

Aerodynamically, the bumble bee shouldn't be able to fly, but the bumble bee doesn't know it, so it goes on flying anyway.

- Mary Kay Ash

ISBN: 978-1-909986-51-0

Introduction

The Entry Level Certificate is a nationally recognised qualification. There are two Entry Level Certificate Science qualifications available.

- Entry Level Certificate (Single Award)
- Entry Level Certificate (Double Award)

The specification is divided into the six components below. Each component has two assessments: a test based on the content in this book set by AQA and an investigation.

Biology

Component 1 - The human body

Component 2 - Environment, evolution and inheritance

Chemistry

Component 3 - Elements, mixtures and compounds

Component 4 - Chemistry in our world

Physics

Component 5 - Energy, forces and the structure of matter

Component 6 - Electricity, magnetism and waves

AF882322

Biology Component 1
The Human Body

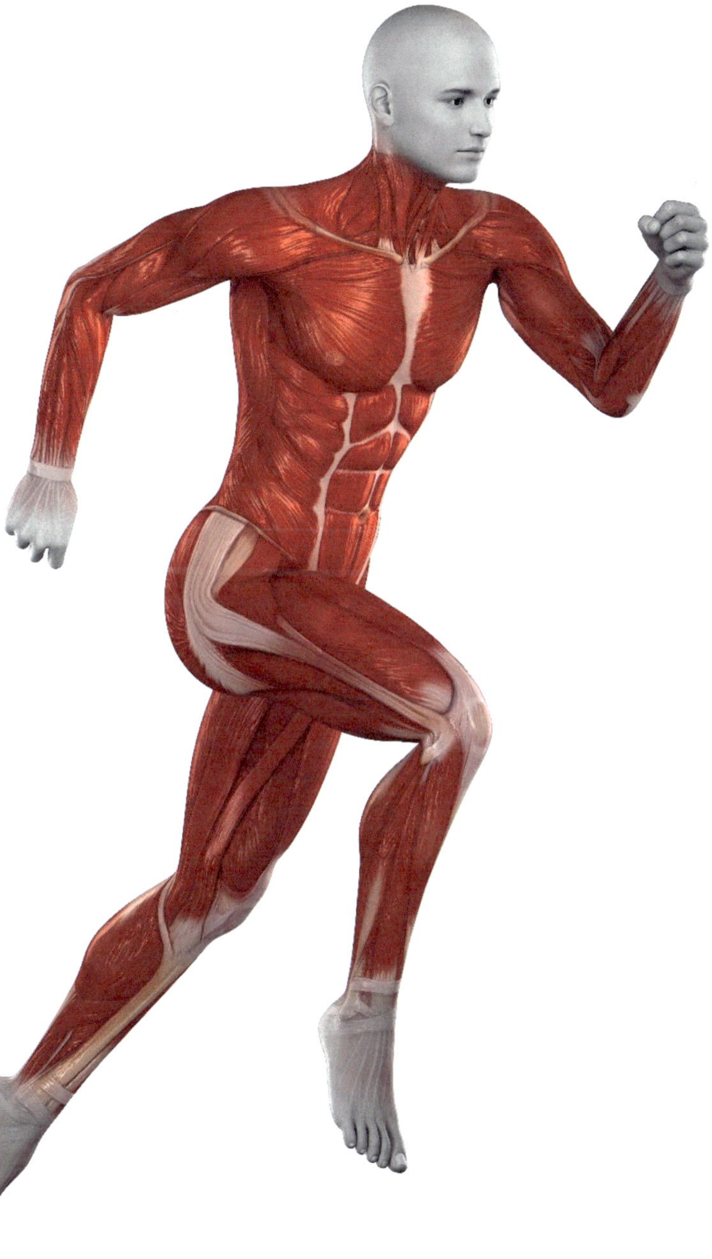

1.1 What is the body made of?

1.1.1 The Cell

At the end of this topic you should know:

- Cells are the basic building blocks of all living organisms.
- Most human cells are like most other animal cells and have the following parts:
    1. nucleus – controls the activities of the cells and contains the genetic material
    2. cytoplasm – where most chemical reactions take place
    3. cell membrane – controls the passage of substances in and out of cells.
- Cells may be specialised to carry out a particular function, e.g. sperm cells, nerve cells and muscle cells.
- The structure of a cell is related to its function (job), and I can explain how.

An organism is any living thing. Organisms include animals, plants, fungi and even bacteria.

The cells of different types of organisms are different.

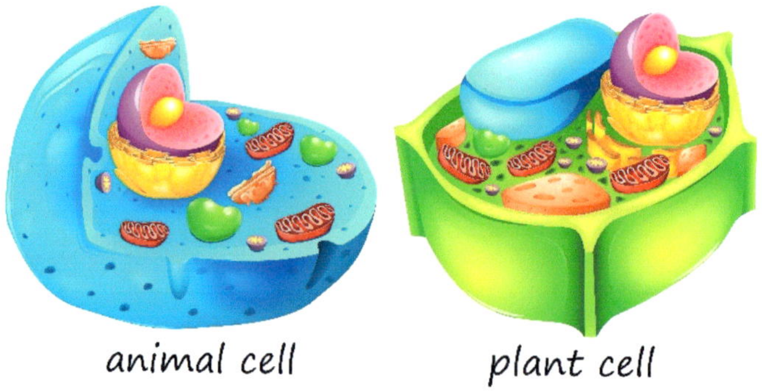

Animal cells including human cells have three main parts:

- Nucleus- this controls the activities of the cells and contains the genetic material.
- Cytoplasm is a jelly-like substance that fills the cell. Most of the chemical reactions occur here.
- Cell membrane controls what goes in and out of the cell.

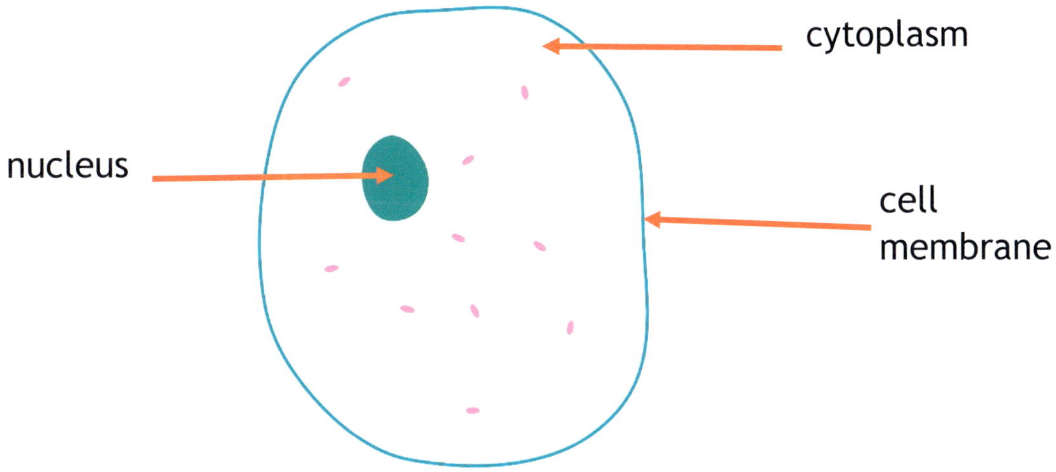

Specialised Cells

Most cells are specialised to carry out a particular function. Different types of cells look different and have different characteristics. These characteristics allow the cells to perform their job.

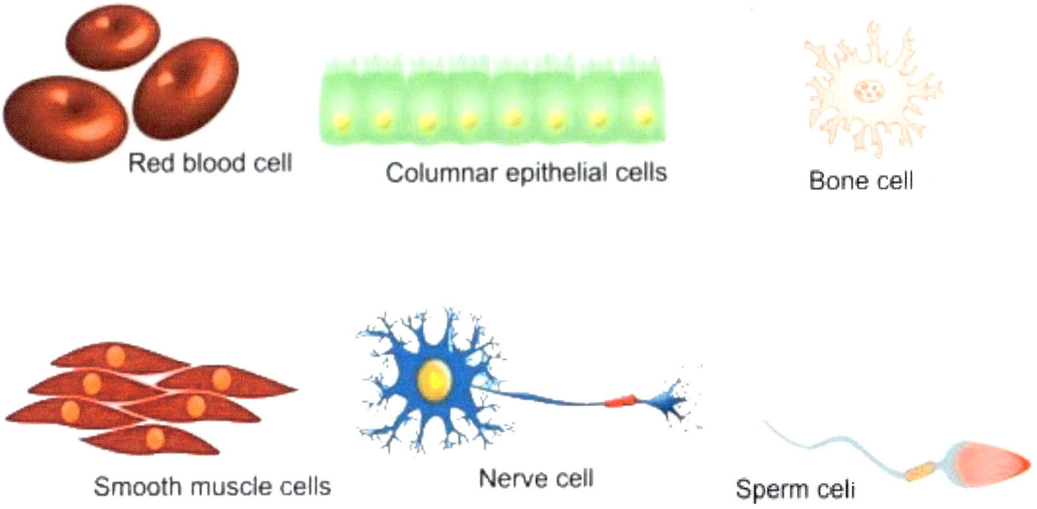

Examples of specialised cells:

Cell	Structure	Function	
Red blood cells	Very small biconcave shape, only human cell without a nucleus.	Carries oxygen around body	No nucleus allows more oxygen to fit in the cell. Very small means it can fit through tiny capillaries.
White blood cells	Flexible shape	ingest pathogens, make antibodies	Flexible shape enables cell to move around and surround pathogen.
Nerve cell	Long, insulated	Sends messages as electrical impulses around body	Insulated means that the messages can travel quickly and don't get mixed up.
Sperm cell	Long tail, nucleus in head.	Provides genes from the father for the new baby.	The long tail enables the sperm to swim to the egg.
Muscle cell	Contains fibres that can contract.	Contracts and relaxes to move bones	Fibres allow muscles to lengthen and shorten.

1.1.2 Organisation

At the end of this topic you should know:

- A tissue is a group of cells with a similar structure and function.
- Have some understanding of the size of cells, tissues, organs and organ systems.
- Organs are groups of tissues performing similar functions.
- Organs are organised into organ systems which work together.
- Be able to identify the position of the major organs in the human body such as the brain, heart, liver, lungs, kidneys and reproductive organs.
- Be able to identify the function of the main organ systems.
- The human circulatory system consists of the heart, which pumps blood around the body (in a dual circulatory system) and blood, which transports oxygen, proteins and other chemical substances around the body.
- Be able to recognise the different types of blood cell from a photograph or diagram.

Cell Organisation

In organisms that are multicellular (have more than one cell) the cells work together.

cell → tissue → organ → organ system → organism

Cell - the building block of all organisms.

Tissue - A group of cells of the same type, working together.

Organ - a group of different tissues working together to perform a particular job.

System - a group of organs that work together to perform a specific task.

Organism - any living thing.

There are many different organ systems in the body including:

System	Organs and tissues include	Function
Circulation	Heart, blood vessels (arteries, veins and capillaries).	Transports nutrient and wastes around the body in the blood.
Respiratory	Trachea (windpipe), lungs, diaphragm.	Taking in oxygen and removing carbon dioxide.
Digestive	Oesophagus (gullet), stomach, large intestine, small intestines.	Breaks down food into simple substances.
Excretory	Kidneys, bladder.	Removes waste.
Nervous	Sense organs, brain, spinal cord, nerves.	Sends messages around body.
Hormone	Hormone glands including thyroid, pancreas, ovaries and testes.	Makes and releases hormones.
Muscular - Skeletal	Bones, cartilage, muscles, tendons and ligaments.	Support and movement.
Reproduction	Uterus, vagina, ovaries, fallopian tube, penis, sperm duct, testes.	Producing offspring.

Human Anatomy

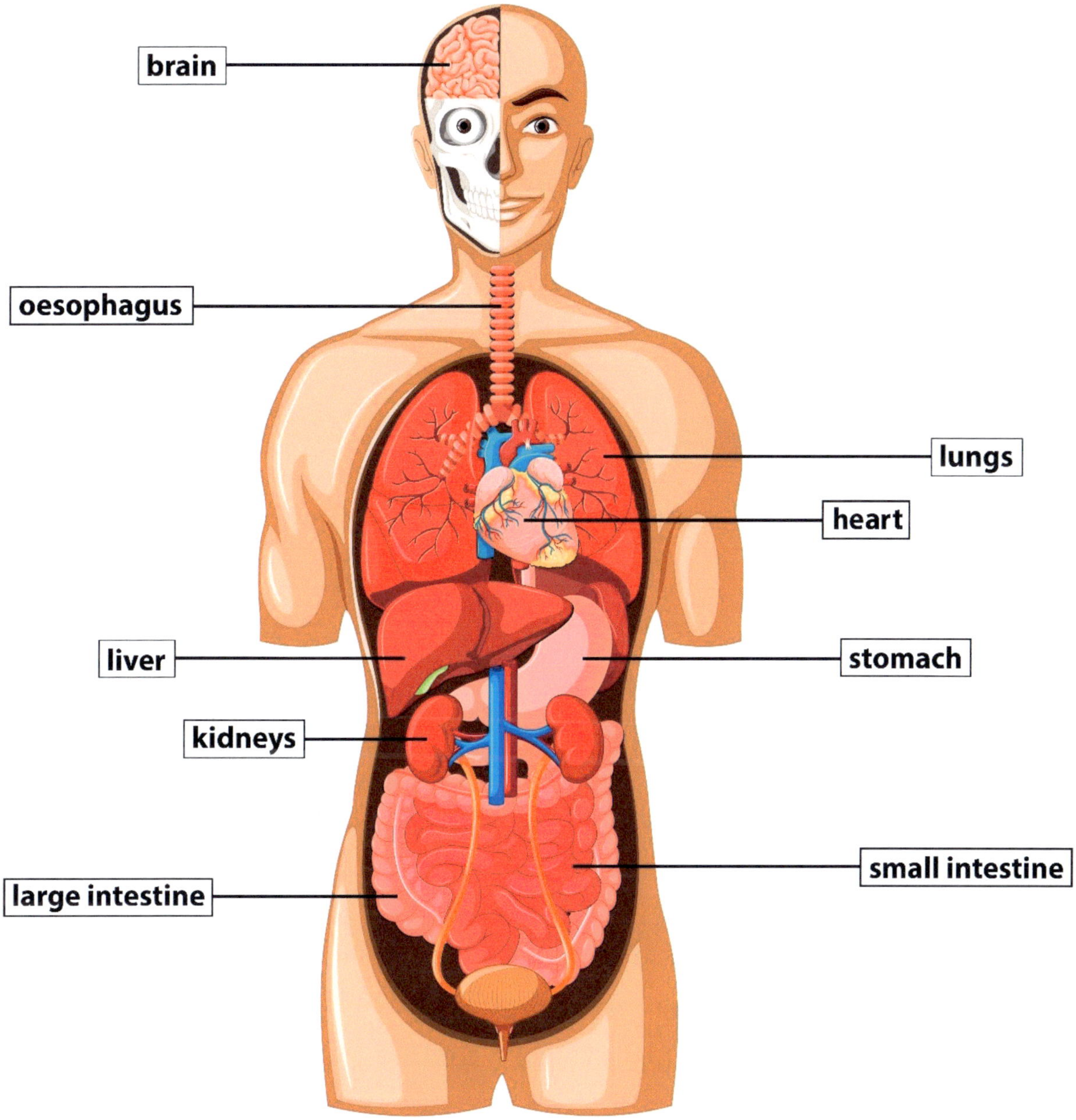

Circulation System

Blood

Blood consists of the following parts:

- Red blood cells
- White blood cells
- Platelets
- Plasma

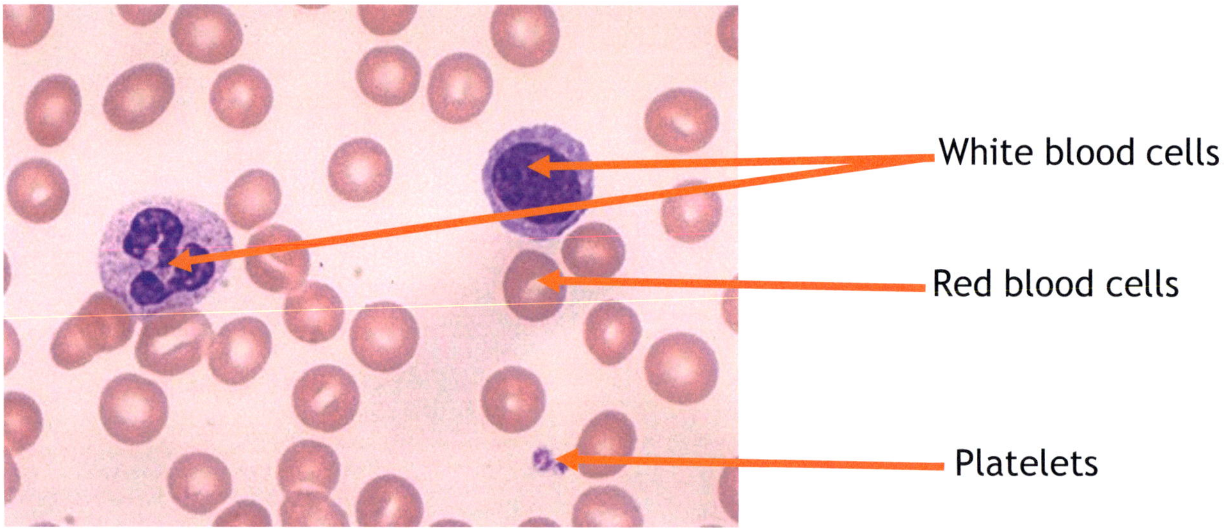

Red blood cells transport oxygen around the body. They:

- contain haemoglobin – a red pigment. Oxygen binds to haemoglobin and is then carried around the body with the haemoglobin.
- do not have a nucleus – leaving more room for oxygen.
- are very small. This allows them to pass through tiny capillaries.

Plasma is a straw-coloured liquid which transports water and dissolved substances such as nutrients, hormones, antibodies, other proteins and waste products around the body.

White blood cells are part of the immune system. They help to protect you against disease.

Platelets help the blood to clot.

Heart

Mammals, including humans, have a double circulation system. This means that the blood passes through the heart twice for each complete circuit of the body.

- The right side takes blood to the lungs
- The left side takes blood to the rest of the body.

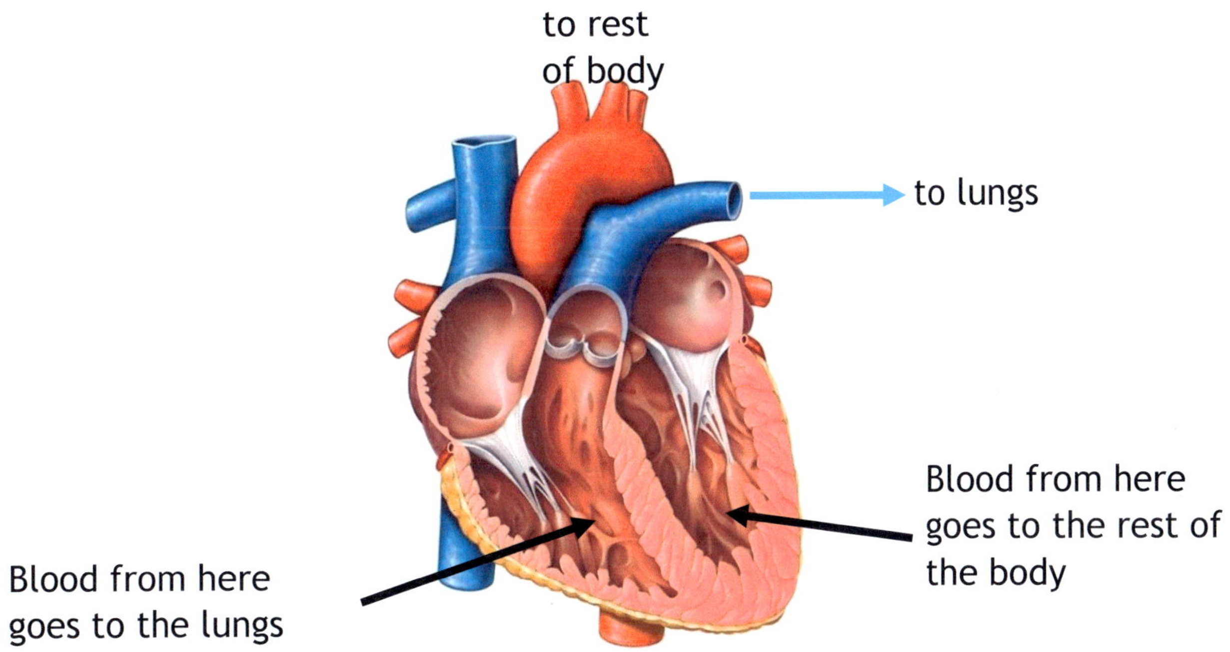

1.1.3 The Digestive System

At the end of this topic you should know:

- The human digestive system contains a variety of organs:
  - salivary glands
  - stomach
  - liver
  - gall bladder
  - pancreas
  - small intestine
  - large intestine.
- Be able to identify the position of these organs on a diagram of the digestive system.
- Enzymes are used to convert food into soluble substances that can be absorbed into the bloodstream.

Digestion

Digestion is the breakdown of food from large insoluble substances to small soluble substances.

There are two types of digestion:

- Mechanical breakdown.
- Physical breakdown.

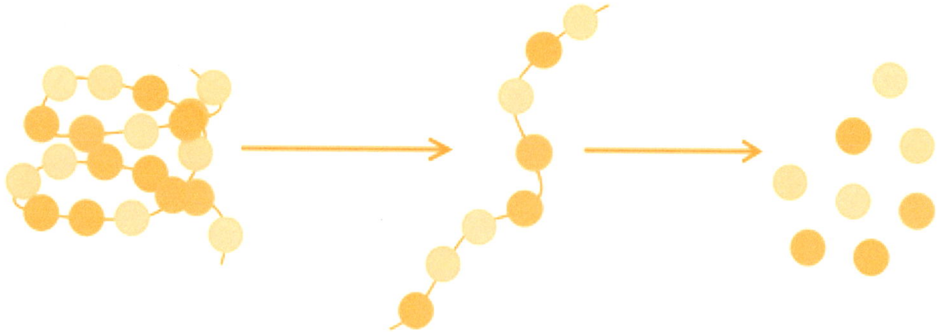

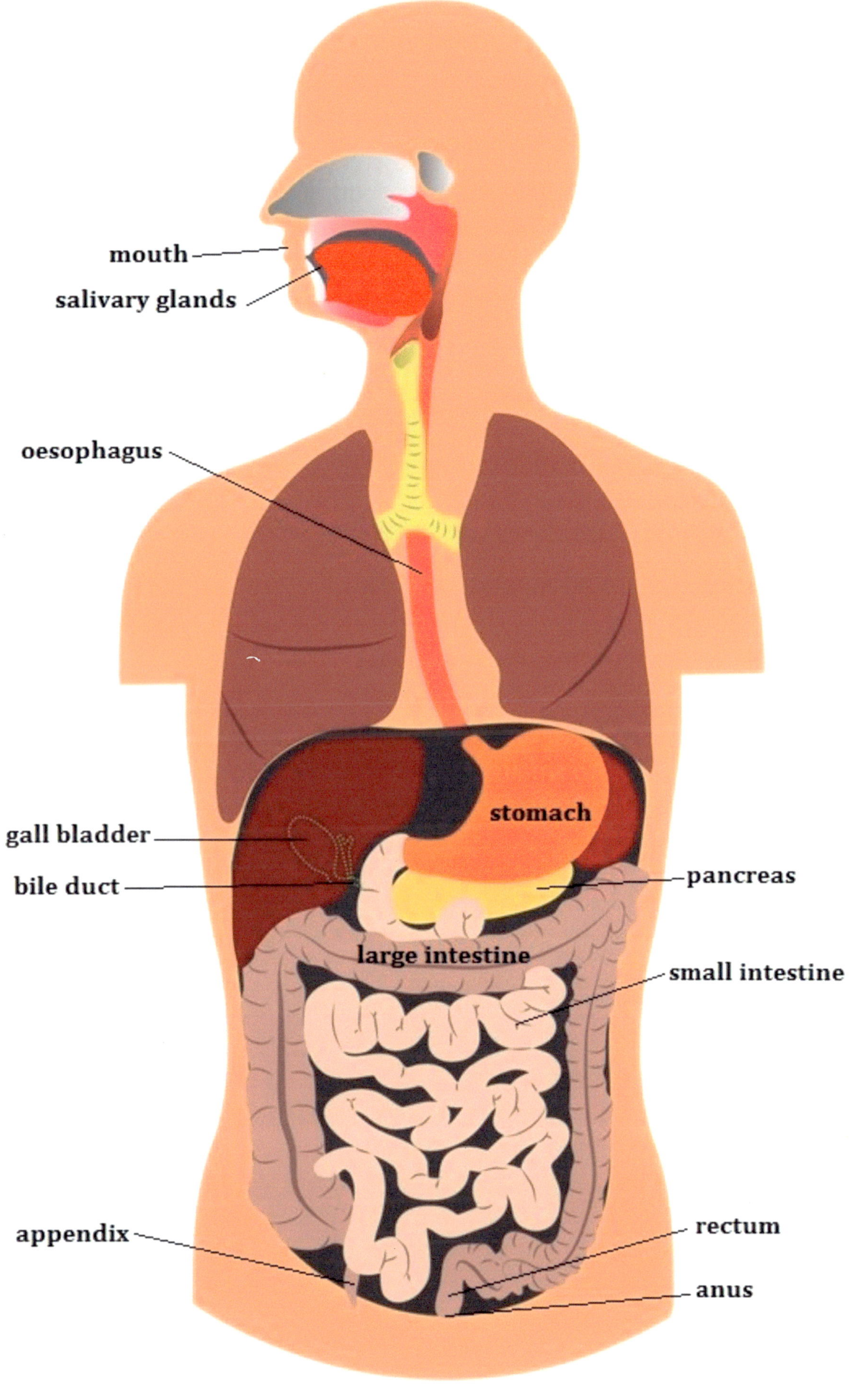
mouth
salivary glands
oesophagus
stomach
gall bladder
bile duct
pancreas
large intestine
small intestine
appendix
rectum
anus

Mouth is where digestion starts. Mechanical digestion with the teeth and chemical digestion with the saliva, made in the salivary glands.
Oesophagus carries food to the stomach.

Stomach is a muscular bag which churns food. It contains hydrochloric acid which kills pathogens.

Liver has many functions including making bile. Bile helps breakdown fats.

Gallbladder stores bile.

Pancreas is a chemical factory that makes enzymes and hormones.

Small intestine is covered with villi, giving a very large surface area. Most food is digested (broken down) here. The small soluble molecules formed are absorbed into the blood.

Large intestine absorbs water and minerals

Enzymes speed up chemical reactions. Digestive enzymes break down large food molecules into smaller ones. These smaller molecules are soluble and are absorbed into the blood stream in the small intestines.

1.2 How the Body Works

1.2.1 Energy

At the end of this topic you should know:

- Respiration releases the energy needed for living processes and is represented by the equation:
- glucose + oxygen → carbon dioxide + water
- (+ energy)
- Glucose comes from the diet and carbon dioxide and oxygen gases are exchanged through the lungs.
- Lifestyle can have an effect on people's health eg diet and exercise are linked to obesity; smoking to cancer; alcohol to liver and brain function.
- A healthy diet contains the right balance of the different foods you need and the right amount of energy.
- People who exercise regularly are usually fitter than people who take little exercise.

Respiration

Respiration is the release of energy by cells. It is ***not*** breathing.

The equation for respiration is:

glucose + oxygen → carbon dioxide + water

Glucose is a sugar and comes from the food eaten.

Oxygen needed for respiration comes from the air you breathe in. The carbon dioxide produced is removed from the body in the air you breathe out. Therefore, the lungs supply oxygen and remove carbon dioxide. This is called gas exchange.

Diet and Exercise

All living things need food. While plants can make their own food, animals need to eat. You need food for two main reasons:

- Energy.
- Provide raw materials for growth and repair.

Some of the things we need energy for are:

- Growth and repair.
- Moving.
- Keeping warm.

Nutrients are substances found in food that are used for growth and repair. Nutrients fall into five main groups:

- carbohydrates.
- proteins.
- fats.
- vitamins and
- minerals.

A balanced diet is one containing the right amount of these nutrients.

Different people need different amounts of food depending on the following factors:

- Gender
- Amount of exercise
- Pregnancy
- Breastfeeding
- Growing
- Sickness
- Body mass

The amount of some nutrients also differ with age. For example, teenagers need more protein because they're growing.

Besides eating a balanced diet to be healthy people need to do some exercise.

It is recommended that adults do:

- at least 150 minutes of moderate aerobic activity such as cycling or brisk walking every week, and
- strength exercises on two or more days a week.

People who do exercise regularly are fitter than those that don't.

If someone has too much food or not enough exercise then they will gain weight. The Body Mass Index, or BMI, is used to determine how much weight someone should be. A BMI of 8.5 to 24.9 means a person is a healthy weight. 25 to 29.9 means they're overweight and above 30 means that they're obese.

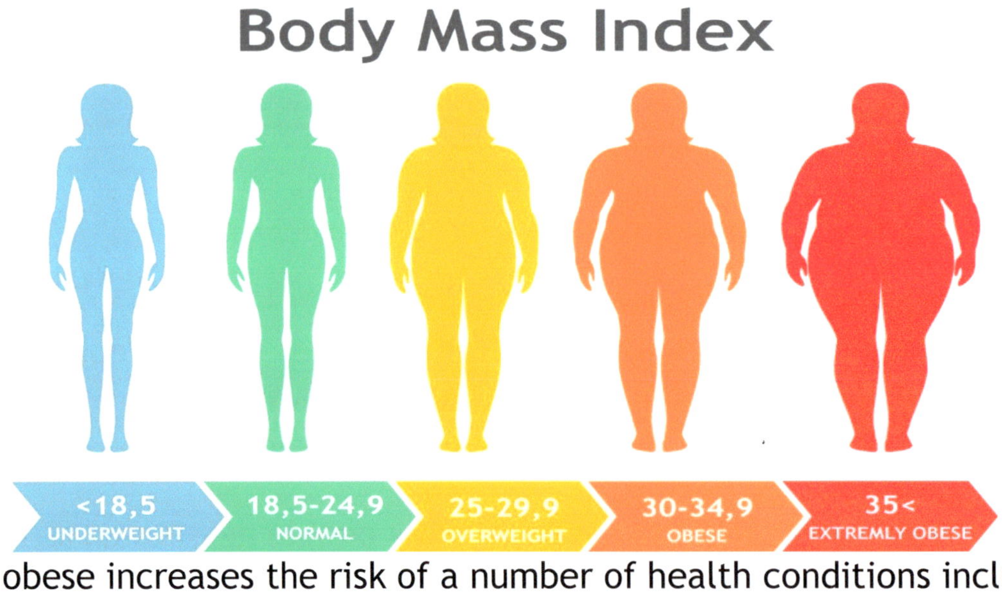

Being obese increases the risk of a number of health conditions including high blood pressure, strokes, heart attacks and type 2 diabetes. Obesity is thought to account for 80-85% of the risk of developing type 2 diabetes, while recent research suggests that obese people are up to 80 times more likely to develop type 2 diabetes than those with a BMI of less than 22.

While the initial symptoms of diabetes tend to be excessive thirst, urination, hunger and extreme tiredness, if untreated serious affects can result including the need for feet to be amputated and damage to the person's vision.

Type 2 diabetes is very different to type one, although the symptoms are the same. Type one diabetes is not caused by diet or exercise but is caused by the body not making enough insulin. Type two diabetes is closely associated to diet and exercise and is caused by the body not being able to use insulin properly.

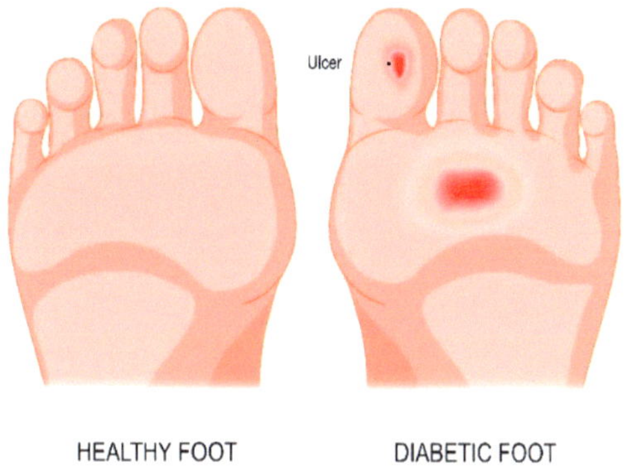

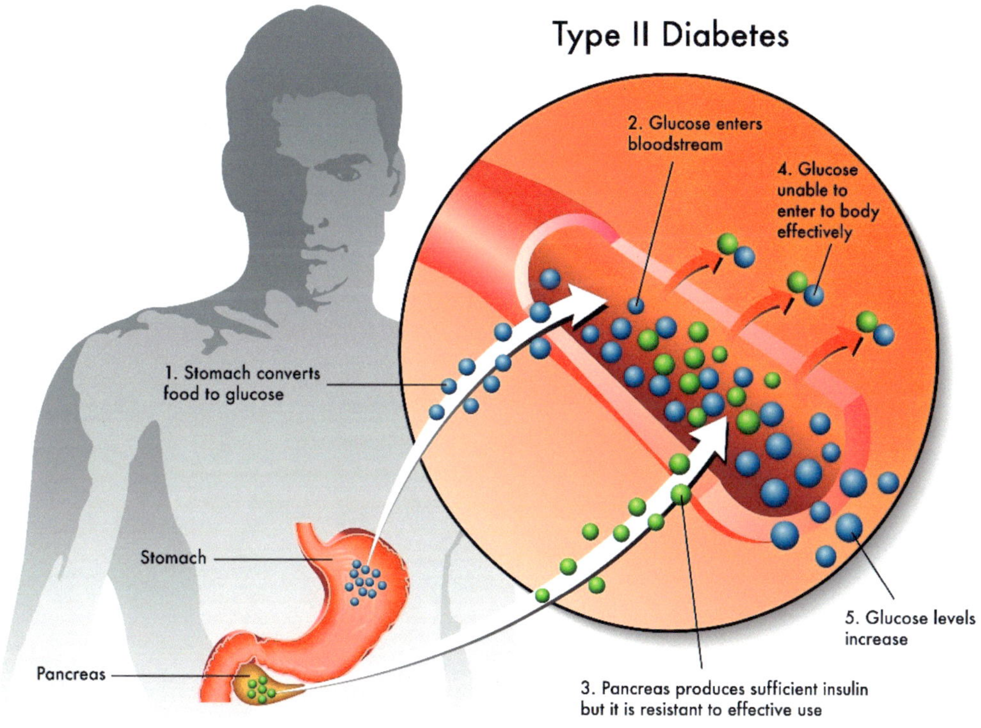

Smoking and Alcohol

Smoking causes severe lung disease such as emphysema, dramatically increases the risk of heart disease and certain cancers such as mouth, oesophagus and lung cancers. The life expectancy of smokers is at least ten years less than for non-smokers. In pregnant woman it is a major cause of still birth as well as medical problems in the baby.

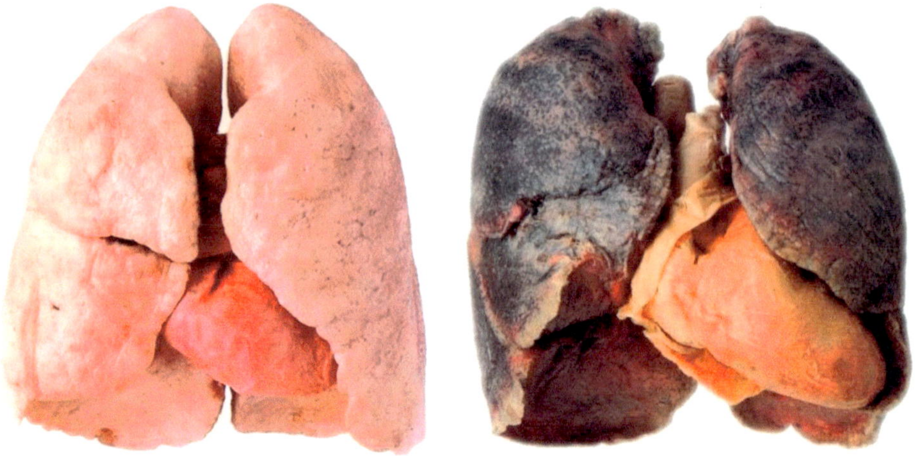

Which lungs come from a smoker?

In the short-term excessive alcohol use can result in impaired judgement. In the long-term excessive alcohol use can cause liver disease and brain damage.

1.3 How the Body Fights Disease

1.3.1 Infectious Disease

At the end of this topic you should know:

- Infectious (communicable) diseases are caused by microorganisms called pathogens.
- These may reproduce rapidly inside the body and may produce poisons (toxins) that make us feel ill.
- Viruses damage cells in which they reproduce.

Pathogens

A pathogen is a disease causing microorganism.

An organism is any living thing, so a microorganism is any very small living thing. Microorganisms are sometimes called microbes. While viruses are non-cellular they are still referred to as microorganisms and pathogens.

Viruses are extremely small, much smaller then bacteria. They consist of a piece of genetic material inside a protein coat. They are not cells. Diseases caused by viruses include:

- Influenza (flu)
- Poliomyelitis
- Measles
- Chicken Pox
- HIV (causes AIDS)

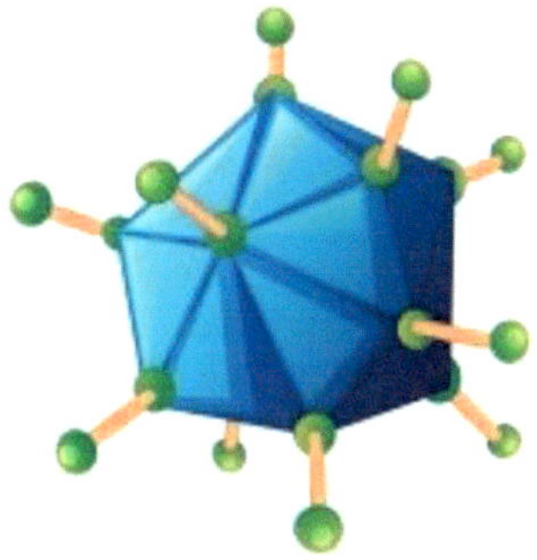

Viruses can only reproduce inside cells. When they do they take over the cell and damage it. Some viruses will fill the cell with copies of itself and then the cell will burst and release the viruses.

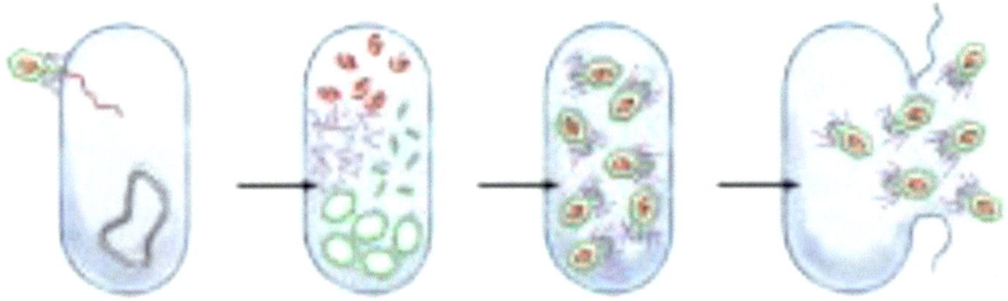

Bacteria are simple cells. They do not have a nucleus. In favourable conditions they can multiply rapidly. They release poisons or toxins that make us feel sick. Diseases caused by bacteria include:

- food poisoning
- cholera
- tetanus
- leprosy
- whooping cough

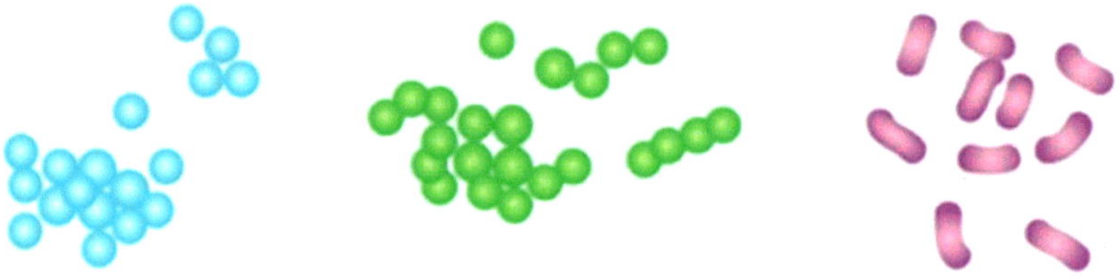

Bacteria release poisons or toxins that make the host organism ill.

Fungi are another type of pathogen. They are bigger than viruses and bacteria. Diseases caused by fungi include:

- Ringworm
- Athletes foot

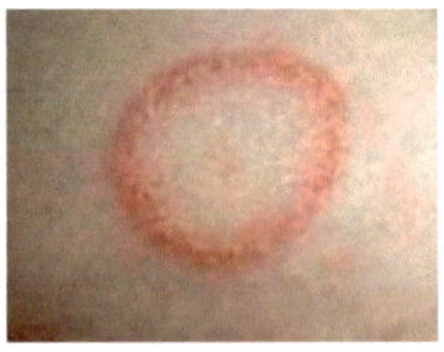

1.3.2 Fighting Infection

At the end of this topic you should know:

- White blood cells help to defend against bacteria by ingesting them.
- Vaccination involves introducing small quantities of dead or inactive forms of a pathogen into the body to stimulate the white blood cells to produce antibodies so that if the same pathogen re-enters the body, antibodies can be produced rapidly.
- Be able to explain the use of vaccination in the prevention of disease.
- Once a pathogen has entered your body, your immune system will work to destroy it.

There are two main types of white blood cell:

- Some ingest the bacteria.
- Some produce proteins called antibodies. Different pathogens need different antibodies.

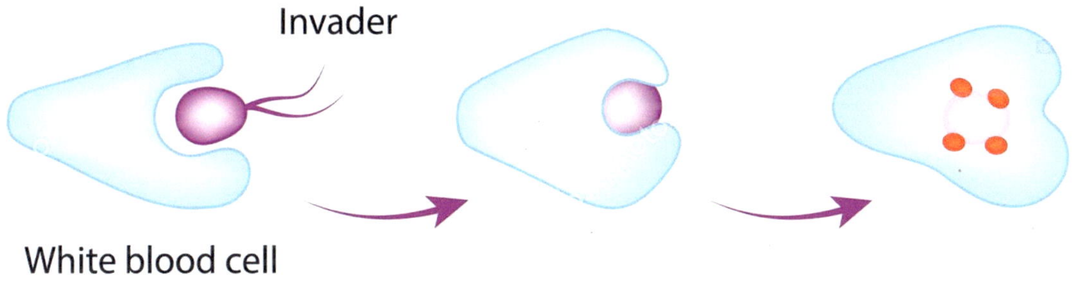

Vaccination

If someone is immune to a disease it means that they don't get the disease even if they come into contact with the pathogen.

In immunisation (also called vaccination) a person is given dead or destroyed pathogens. The immune system will react to these pathogens and form antibodies. However, they will not become sick as the pathogens are dead or unable to divide.

The first time the body makes antibodies for a pathogen it is slow because it needs to produce the exact shape, however, if it needs to make the same antibodies later it can do so much quicker.

Therefore, if a person has been vaccinated, when they get the real disease their body can make lots of the antibodies rapidly and destroy the pathogen before it has a chance to make them sick.

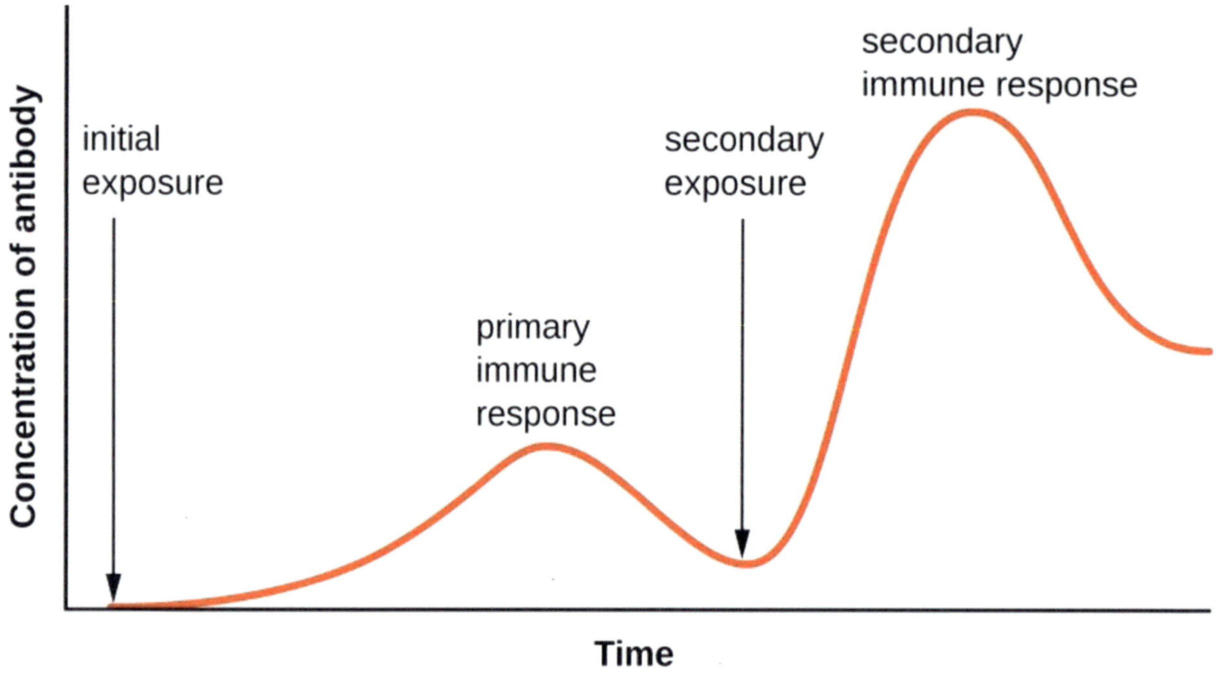

1.3.3 Using Drugs to Fight Disease

At the end of this topic you should know:

- Medical drugs are developed and tested before being used to relieve illness or disease.
- Drugs change the chemical processes in people's bodies.
- People may become dependent or addicted to the drugs and suffer withdrawal symptoms without them.
- Antibiotics, including penicillin, are medicines that help to cure bacterial disease by killing infective bacteria inside the body, but cannot be used to kill viruses.

What are Drugs?

A drug is anything that changes how the body works and is not food.

Effects of Drugs

In medicine, a **side effect** of a drug is an effect that is not the one that the drug was given for. Side effects are normally negative and often involve tiredness, nausea or headaches.

When a person has been on a drug for a while the person can become **tolerant** to the drug. Tolerance is a person's reduced reaction to a drug as their body adapts to the drug. This means they need more of the drug to have the same effect and eventually the drug becomes ineffective.

People may become **addicted** to drugs. This means that if they stop taking the drug, they will suffer **withdrawal symptoms**. An example of this is someone who drinks a lot of coffee, if they stop drinking coffee, they will suffer headaches and crave a cup.

Antibiotics

Antibiotics are drugs that kill bacteria or stop them from reproducing. They do not kill viruses. The first antibiotic discovered was penicillin which works by stopping bacteria from making a cell wall. Since viruses are not cells and do not have a cell wall this has no effect on viruses.

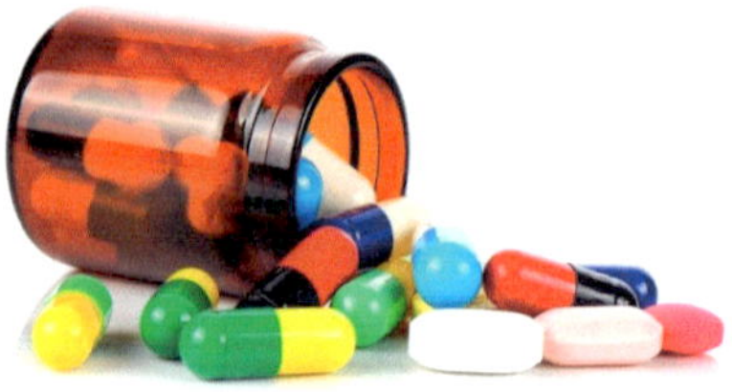

Development of Drugs

Drug companies have to follow strict procedures before they can release new drugs. There are three main steps to testing potential drugs:

1. Testing the drugs on cells or tissues in the lab.
2. Then the drug is tested on animal models to see how the drug works in a whole organism.
3. The last step is the clinical trials. Each drug will go through a series of clinical trials.

In all of the stages a placebo is used. A placebo is the same as the drug being tested in every way except for the active ingredient.

In clinical trials the patients are put into at least two groups. One group will be given the drug being tested; the other group will be given the placebo.

It is important in clinical trials that neither the doctor nor the patient know if they are taking the drug or the placebo. These trials are double blind. This is important because it reduces bias.

The drug trials will not only compare the drug with the placebo to check that it is effective, but it will also look at what side effects the drug has. What side effects are acceptable depends on how serious the disease is.

1.4 How the Body is Coordinated

1.4.1 Nervous System

At the end of this topic you should know:

- The human body has automatic control systems, which may involve nervous responses or chemical responses coordinated by hormones.
- Reflex actions are automatic and rapid.
- Examples include the response of the pupil in the eyes to bright light, and the knee jerk reaction.

Communication in the body

The body has two communication systems:

- The nervous system – electrical impulses travelling along nerves.
- The hormone system – chemical messengers that travel through the blood.

The nervous system acts in a fraction of a second for a short period of time, while the hormone system takes much longer to respond but the effect occurs for much longer, often hours or even days.

Reflex actions

Some actions of the nervous system are controlled by our thoughts. However, other actions are automatic. They can occur without the message needing to go through the brain first. These very fast, automatic responses are called reflex actions.

There are a number of reflex actions including:

- The pupil of the eye getting smaller in light and larger in the dark. This prevents too much light entering and damaging our eyes but allows us to see in darker conditions.
- The knee jerk reaction.

1.4.2 Hormone System

At the end of this topic you should know:

1. Hormones are secreted by glands and are transported to their target organs by the bloodstream.
2. Several hormones are involved in the menstrual cycle of a woman, including some that are involved in promoting the release of an egg.
3. Should be familiar with a diagram of the menstrual cycle.

Hormone System

A hormone is a chemical messenger. They are made and secreted into the blood, by glands. The hormone then travels through the blood to the target organs. The hormone connects with or enter the target organ where it has its effect.

An example of a hormone is insulin. Insulin is made in the pancreas (the gland) which secretes it into the blood. Insulin controls the amount of sugar in the blood.

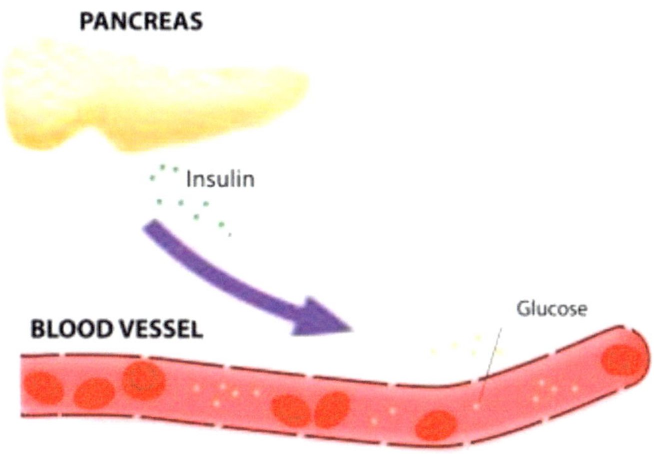

Another hormone is adrenaline. Adrenaline is produced in the adrenal glands situated just above the kidneys. Adrenaline causes the "fight or flight" response. It is also the chemical in a "EpiPen" that is used by people with severe allergies.

Menstrual cycle

After puberty females have a monthly cycle caused the menstrual cycle. The cycle is different in different females. The menstrual cycle is controlled by hormones.

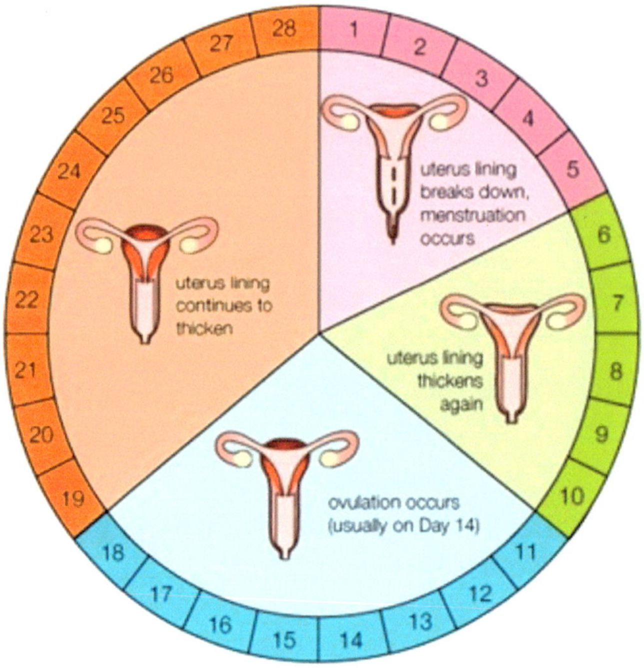

The average cycle is 28 days. On around day 14 an egg is released from the ovary. This is called ovulation. If the egg is not fertilised, then the egg passes out of the body with the lining of the uterus and some blood. Day one of the cycle is the start of bleeding. This bleeding is called menstruation or a period.

The thickness of the uterine lining during a cycle of the menstruation cycle.

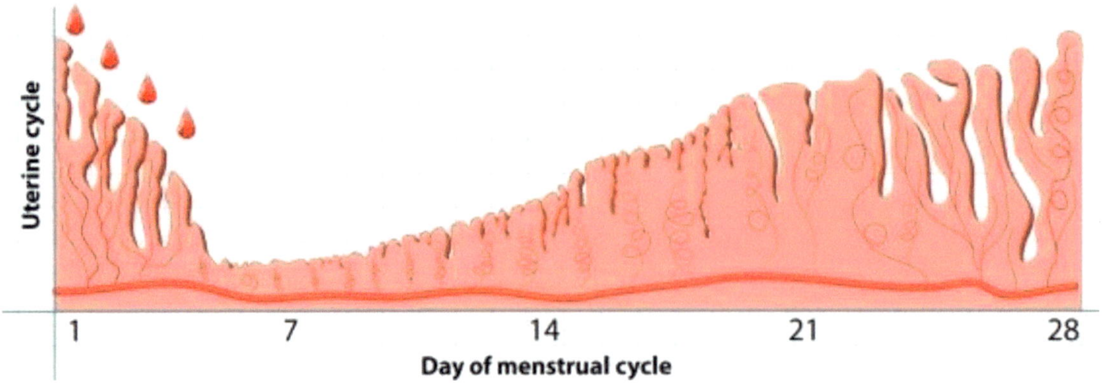

Periods stop when a female reaches menopause. After menopause a female is no longer able to have children.

1.4.3 Fertility

At the end of this topic you should know:

- The uses of hormones in controlling fertility include:
  - giving oral contraceptives that contain hormones to inhibit eggs from maturing
  - giving 'fertility drugs' to stimulate eggs to mature.
- Evaluate the benefits of, and the problems that may arise from, the use of hormones to control fertility.

Controlling fertility

It is possible to stop women becoming pregnant by using the contraceptive pill. The contraceptive pill contains hormones which stop the woman from releasing an egg.

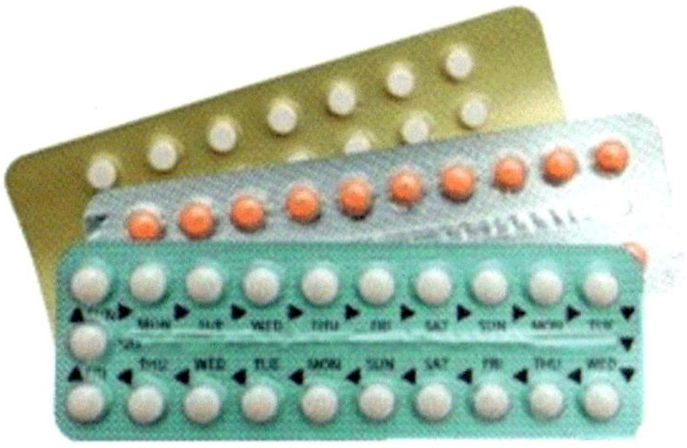

It is also possible to use hormones to help a woman who can't have children become pregnant. These drugs are called "fertility drugs". These drugs cause the woman's eggs to mature and be released.

Advantages and Disadvantages

The use of the contraceptive pill has the following advantages:

- reliable
- effective
- readily available

However, they have the following disadvantages:

- Need to be taken daily
- Against some people's beliefs.

Biology Component 2

Environment, Evolution and Inheritance

2.1 Feeding Relationships between Organisms

2.1.1 Photosynthesis

At the end of this topic you should know:

- Radiation from the sun is the source of energy for living organisms.
- Green plants and algae absorb light from the sun and use it to make glucose.
    - This process is called photosynthesis
    - The reaction is: carbon dioxide + water → glucose + oxygen

Light from the sun provides the energy for most organisms.

Green plants and algae make their own food in a process called photosynthesis. Because they make their own food, they are called producers. To do this they take in light energy from the sun. They also need carbon dioxide and water.

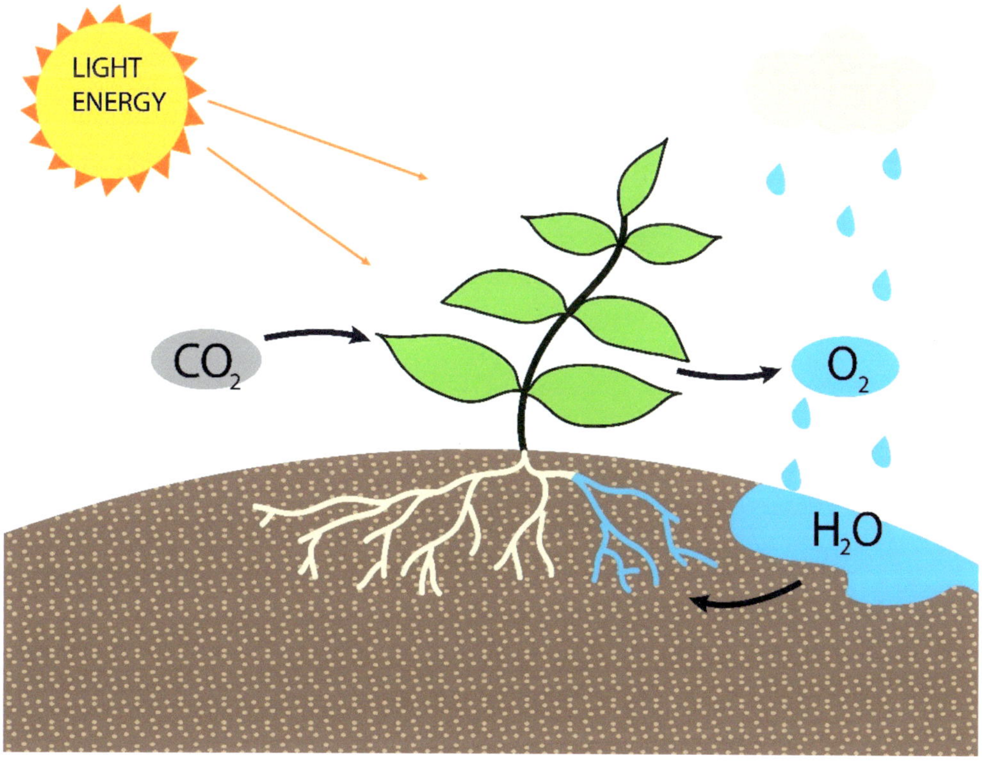

The plant or alga then combines the carbon dioxide and water to make glucose and oxygen.

The word equation is:

carbon dioxide + water → glucose and oxygen.

Remember, when writing the word equation:

- Carbon dioxide and water can be written either way around.
- Glucose and oxygen can be written either way around.

The important thing is that you start with the carbon dioxide and water and finish with the glucose and oxygen.

Remember, the energy for this process comes from sunlight.

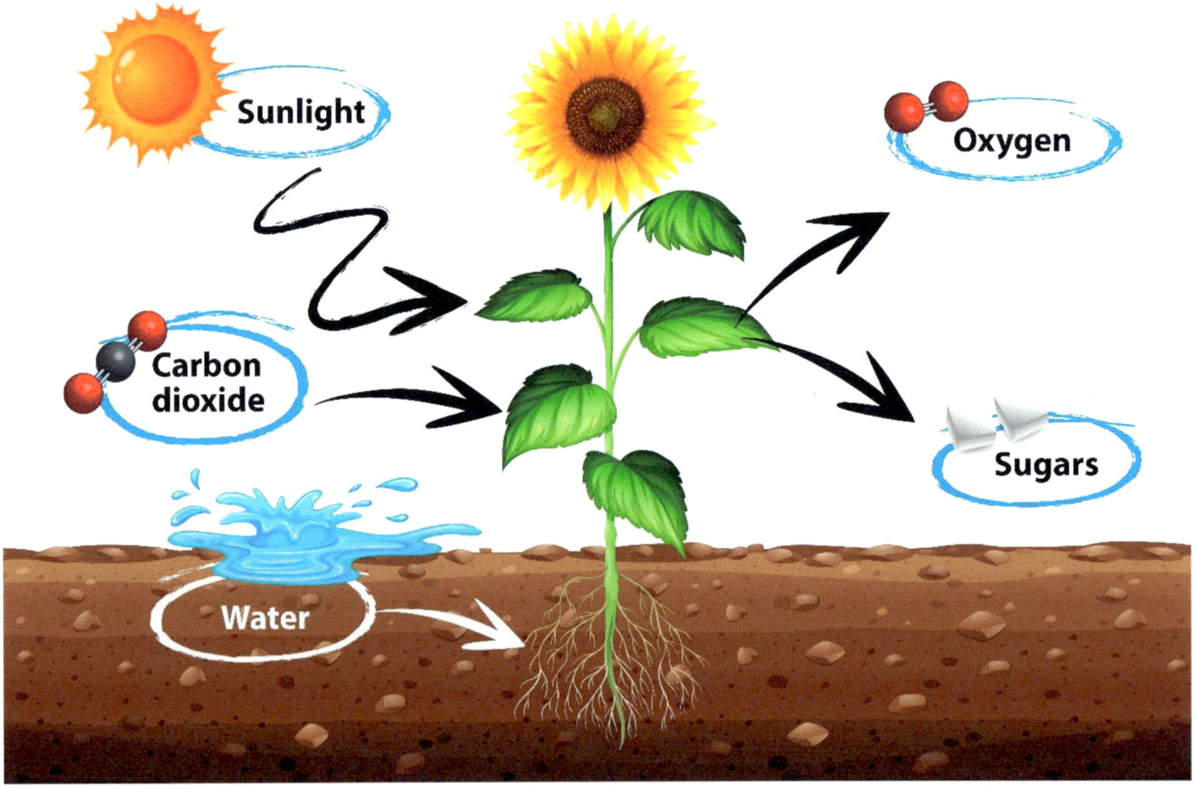

2.1.2 Adaptations

At the end of this topic you should know:

- Animals and plants may be adapted for survival in the conditions where they normally live.
- You should be able to explain how organisms are adapted to live in their natural environment given appropriate information in image or prose format. Examples may include polar bears in the Arctic, or cacti in deserts.

A couple of definitions:

Organism - any living thing; such as a plant or an animal.

Habitat - where an organism lives.

An adaptation is any characteristic that an organism has, that allows it to survive in a particular habitat.

General Adaptations

Some general adaptations include:

- Camouflage - the organism is the same colour or pattern as the surroundings making them difficult to see.
- Fast - able to get away from predators or catch prey.

This is a Vietnamese mossy frog:

This insect is called the Ghost Mantis.

Would you notice this frog, if you walked past the tree?

Or would you walk past this Leaf-Tailed Gecko?

Cold climates

Animals in cold climates, such as the polar bear, will have a range of adaptations, including:

- Large feet – to spread out the weight and stop the animal sinking into the snow.
- Thick fur – to insulate the animal and keep it warm.
- Small ears – to reduce heat loss.
- Huddling together – keeps them warm, reducing heat loss.

Notice the white colouring, large feet and thick fur of the Polar Bear.

The Emperor Penguins below will take it in turns to be on the outside and the inside of the group.

Desert animals

Some animals are adapted to cope with the harsh desert conditions, including the heat and lack of water.

An example of an animal adapted to the desert is the camel. The camel is adapted by:

- Large flat feet – spreads their weight, stopping them from sinking into the sand.
- Two rows of eyelashes – stop sand from getting in.
- Nostrils that close – stops sand from getting in.
- Hump – stores fat.
- Produces little urine or sweat – reduces water loss.

Desert plants

Plants also have adaptations. The cactus is adapted to survive with very little water. The adaptations of the cactus include:

- Waxy surface – prevents water loss.
- Succulent stem – stores water
- Spines instead of leaves – less water lost and stops animals eating it.
- Widespread roots – When it rains, they can collect water very quickly over a large area.

2.1.3 Food chains and Food webs

At the end of this topic you should know:

- Feeding relationships within a community can be represented by a food chain. All food chains begin with a producer.
- A food web can be used to understand the interdependence of species within an ecosystem in terms of food resources.

The feeding relationship between producers and consumers can be written down in a series of steps called a food chain. For example

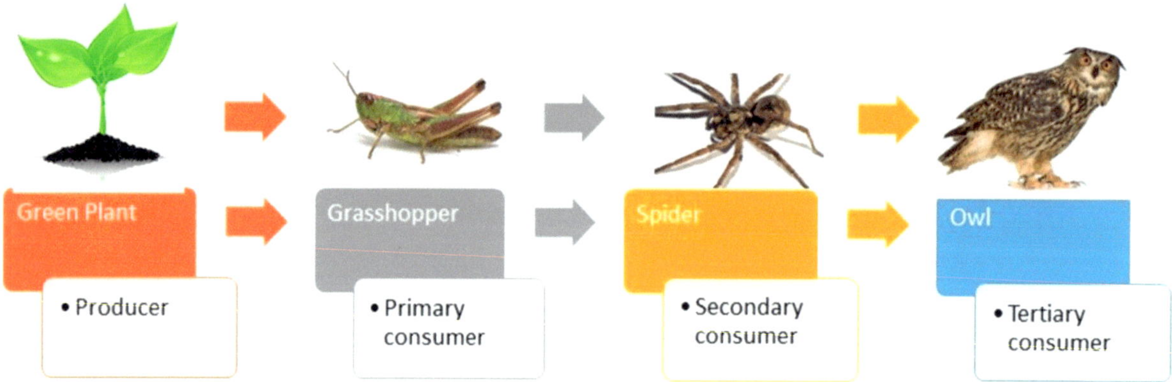

Food chains are always written with the arrows showing the direction of energy transfer. Food chains always start with a producer (an organism that makes its own food). This is often but not always a plant.

In the above food chain, the grasshopper eats the green plant, the spider eats the grasshopper and the owl eats the spider.

A food chain does not give the complete picture, as a spider does not only eat grasshoppers, nor is it the only organism which eats grasshoppers. Similarly, an owl does not dine solely on spiders. When many food chains interlock a food-web is created.

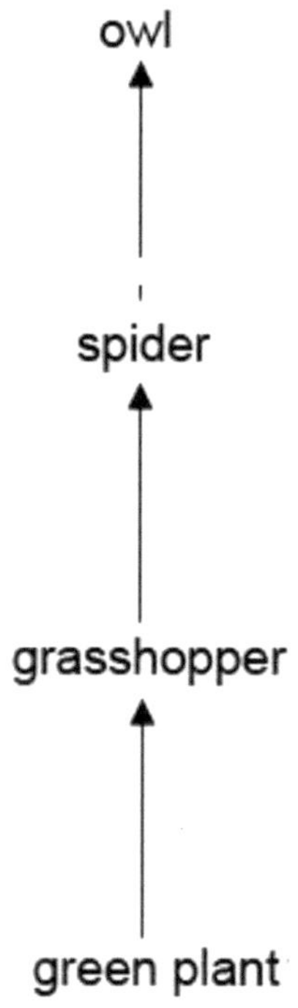

A Food Chain

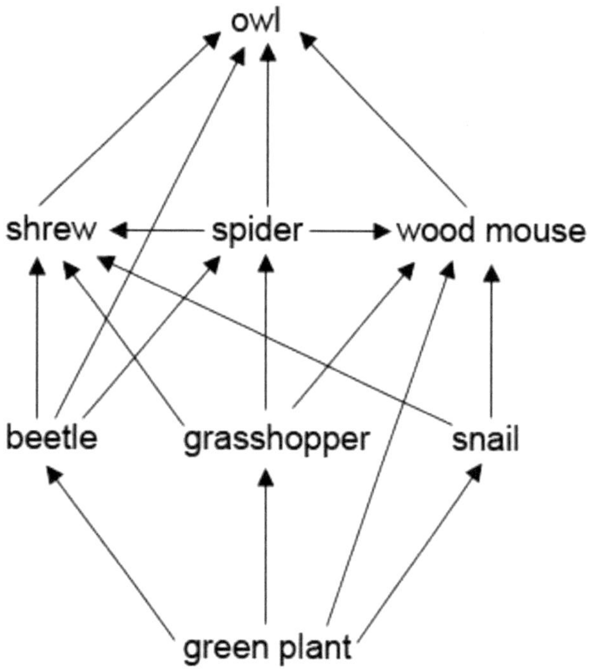

A Food Web

A food web represents the flow of energy between organisms, or the feeding relationship between organisms, within a community.

When one type of organism in a food web increases or decreases, it affects every other organism in the food web.

If the number of grasshoppers decrease, the spiders will eat more beetles. If the spider decreases the owl will eat more shrews, beetles and wood mice.

As more wood mice and shrews are being eaten there will be more snails. So, even though the snail is not directly in a food chain with the spider, it is still affected by it.

2.1.4 Nutrient Recycling

At the end of this topic you should know:

- All materials in the living world are recycled to provide the building blocks for future organisms.
- Decay of dead plants and animals by microorganisms returns carbon to the atmosphere as carbon dioxide to be used by plants in photosynthesis.

The different elements that make up plants and animals are not created or destroyed but are constantly cycled through the environment.

While the plants and animals are alive, they take in nutrients from the environment. Plants take in nutrients from the soil and bring in carbon dioxide through their leaves. Animals eat other organisms for food.

This means that all material in ecosystems is recycled and used again and again to make living things.

When these plants and animals die, they decay. Microorganisms, such as bacteria, break down the dead organism returning the nutrients back to the environment. Waste produced by animals and plants also decay in the same way.

The Carbon Cycle

The main steps of the carbon cycle are:

- Carbon is taken in by plants as carbon dioxide.
- The plant converts this carbon into glucose, a type of sugar.
- The plant changes this glucose into many things, such as cellulase and oils and it becomes a part of the plant.
- The plant is then eaten by animals.
- When the plant and animal dies, they are decayed by microorganisms.
- The process of decay releases the carbon back into the atmosphere as carbon dioxide.

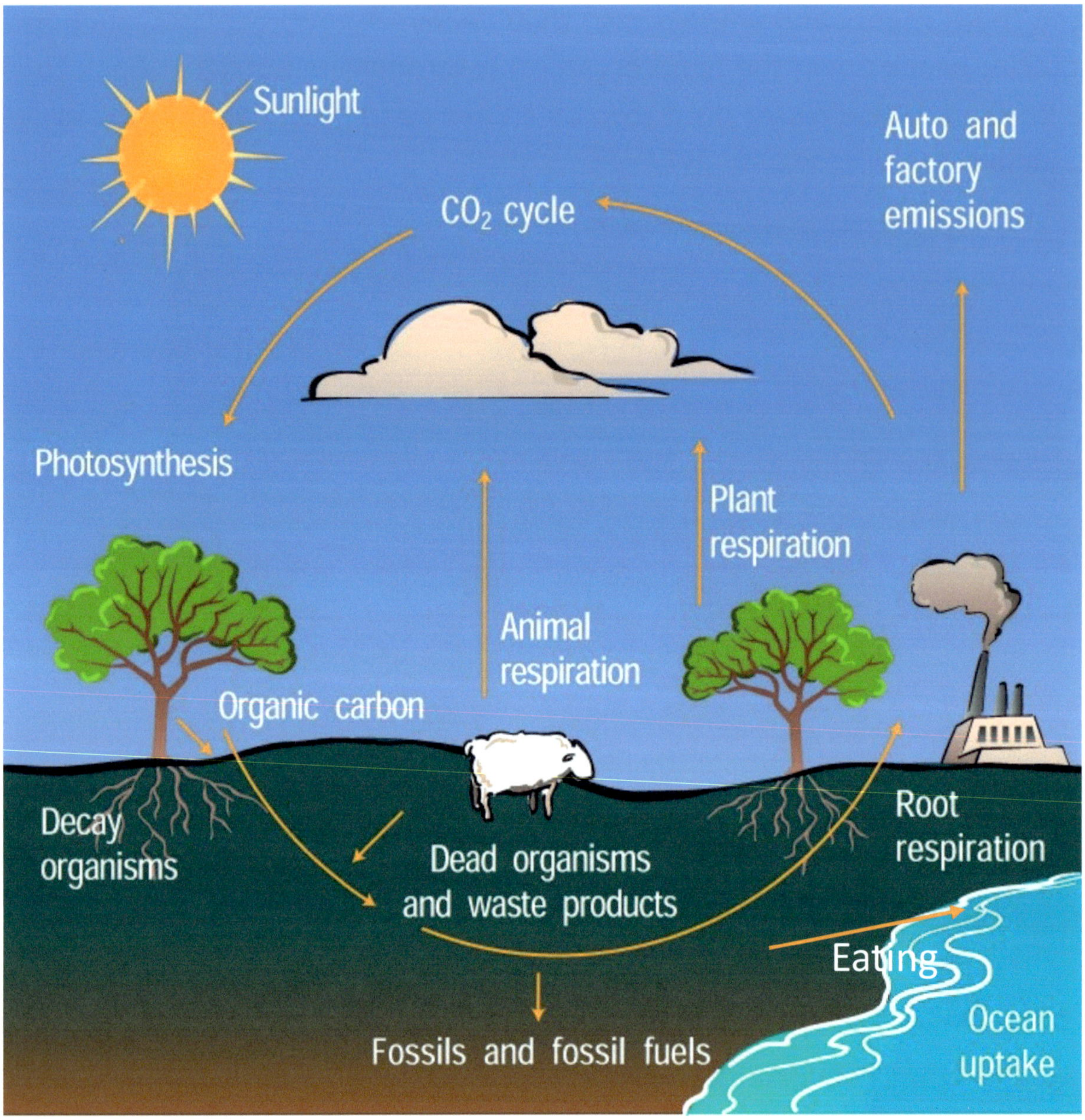

So, plants and algae take carbon out of the atmosphere. Decay returns that carbon back to the atmosphere as carbon dioxide.

2.2 What Determines Where Particular Species Live?

2.2.1 Competition

At the end of this topic you should know:

- Plants often compete with each other for light and space, and for water and nutrients from the soil.
- Animals often compete with each other for food, mates and territory.

Competition happens both within a species and between different species.

Plants compete for:

- Light
- Space
- Water
- Nutrients

Animals compete for:

- Territory (space)
- Food
- Water
- Mates

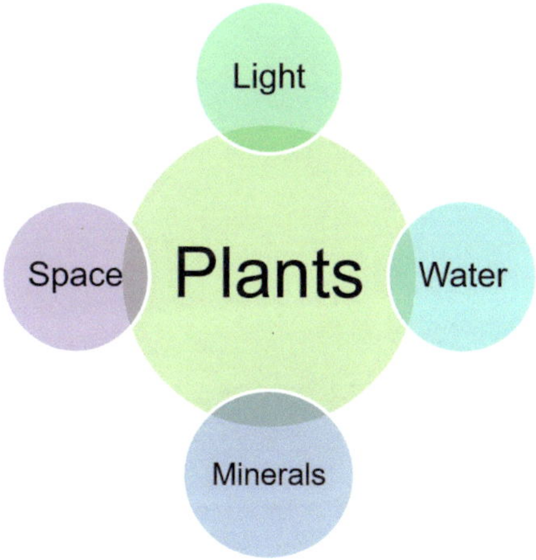

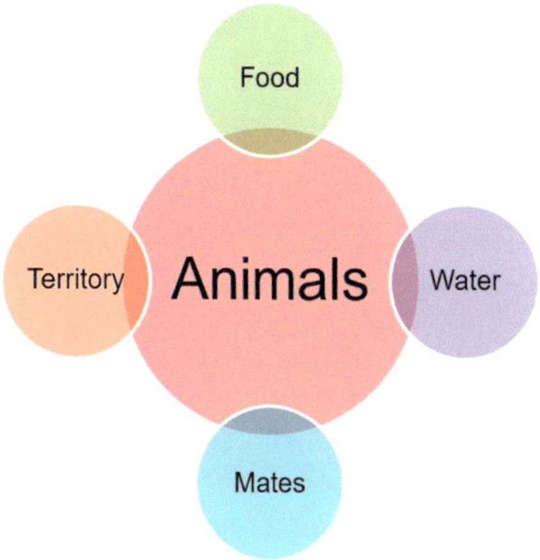

2.2.2 Environmental Changes

At the end of this topic you should know:

- Animals and plants are subjected to environmental changes. Such changes may be caused by non-living or living factors.
- An example of a non- living factor is a change in the average temperature or rainfall.
- An example of a living factor is the introduction of a competitor or predator.

Changes in both living and non-living factors will change the environment.

Non-living factors include:

- Average temperature – temperature affects which animals and plants are able to survive as well as the number of them.
- Rainfall – all organisms need water, but some are adapted to survive with less available water. Drought can have a huge effect on both plants and animals.

Living factors include:

- Predators – for example a new predator may dramatically reduce prey that do not have defences against them.
- Prey – a decrease in prey will cause a decrease in their predators as there is less food. Similarly, an increase in prey will cause an increase in predators as there is more food.
- Competitors – more competitors mean more competition for food and space.
- Trampling – some plants can survive being trampled by humans or other animals, many cannot.

The number of red squirrels in the UK is decreasing due to the competition from the grey squirrel.

2.2.2 Pollution

At the end of this topic you should know:

- Pollution of the environment can occur:
    - in water, from sewage, fertiliser or toxic chemicals
    - in air, from smoke and gases such as sulfur dioxide which contributes to acid rain
    - on land, from landfill and from toxic chemicals such as pesticides and herbicides, which may be washed from land into water.
- The rapid growth in human population means that more resources are used, and more waste is produced.

The human population is growing rapidly. The larger the human population the more resources that are needed and the more waste produced.

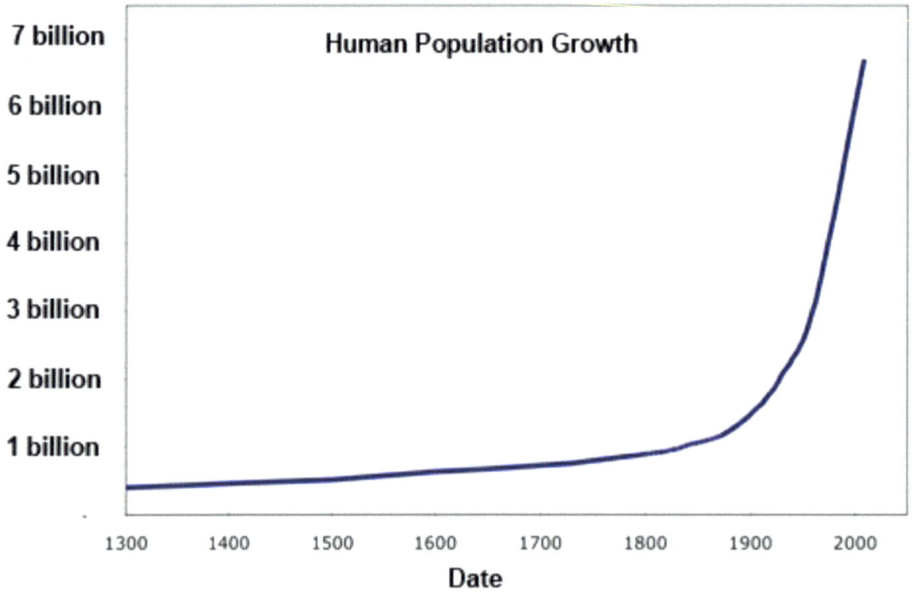

One way that a greater population increases pollution, is there is more transport.

Air Pollution

Burning fossil fuels produces sulfur dioxide. Cars and other vehicles produce nitrogen oxides. The sulfur oxides and nitrogen oxides are gases that dissolve in water and forms acid rain.

Acid rain affects nearly everything. Plants, soil, trees, buildings and even statues can be damaged.

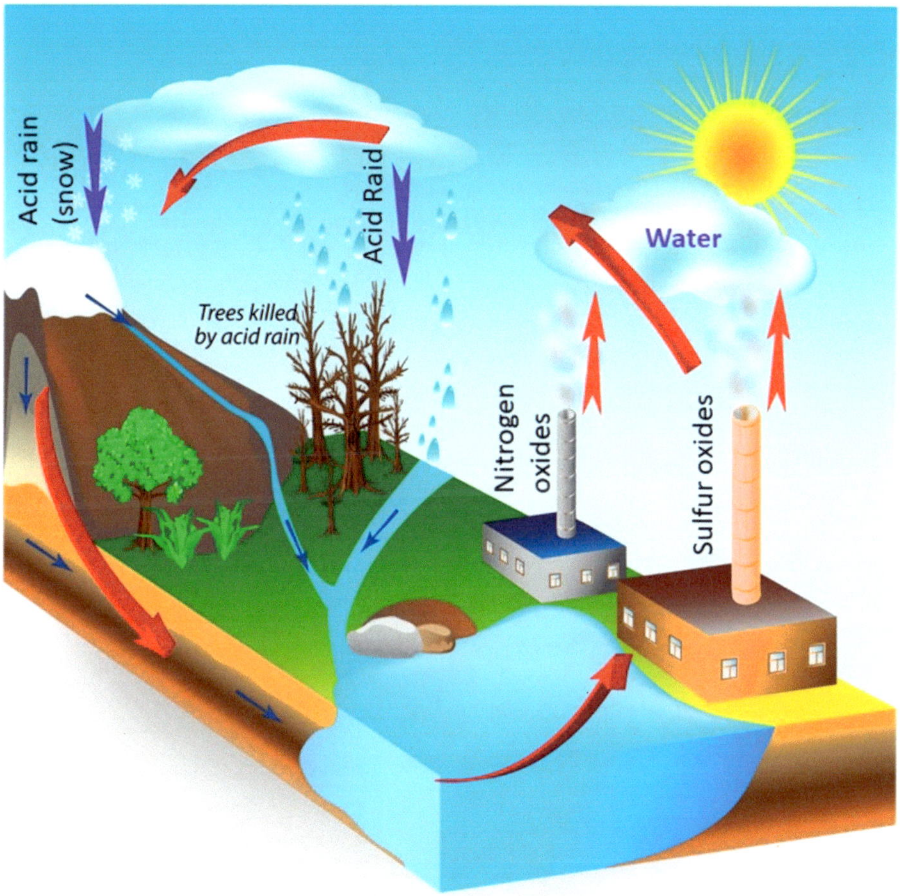

Waste from factories can contain toxic chemicals. Toxic means poisonous and these chemicals can harm or even kill living things.

Land Pollution

Farmers put fertilisers on crops to help them grow.

Farmers and gardeners also use pesticides and herbicides.

Pesticides kill pests and herbicides kill weeds.

These chemicals can be toxic and hurt or even kill living things.

Landfill is a major problem and the more people the greater the amount of waste that will go into landfill. Landfill often contains toxic chemicals that spread into the soil making the ground unusable. As the waste breaks down harmful gases are also produced.

Water Pollution

The fertilisers that farmers put on the crops can get into water. The fertilisers cause algae to grow and cover the surface of the water. This is called an algal bloom. This stops sunlight from reaching the plants and animals below. It also causes the oxygen in the water to drop.

Waste from humans is called sewage. Sewage when it gets into rivers can cause lots of diseases such as cholera. Diseases caused by water pollution cause at least 1.5 million deaths each year, mostly of young children.

Some factories also release waste that contains toxins into rivers. This can cause disease in living organisms or even kill some living things.

2.3 How Life has Developed on Earth?

2.3.1 Natural and Artificial Selection

At the end of this topic you should know:

- Darwin's theory of evolution states that all species of living things have evolved from simple life forms that first developed more than three billion years ago.
- In natural selection, individuals with characteristics most suited to their environment are most likely to survive to breed successfully.
- Artificial selection (selective breeding) is the process by which humans breed plants and animals for particular genetic traits.

Evolution is how organisms change over time.

Evolution can refer to either:

- small changes like bacteria developing antibiotic resistance or Peppered Moth changing colour.
- Big changes where a new species is formed.

Scientists have been able to observe the small changes. We cannot see the big changes because they occur over a very long period of time. Evolution referring to these big changes is a theory. It is the best theory that Scientists have to explain how the different organisms came to be. You don't have to agree with it, but you do have to be able to explain it.

Charles Darwin travelled to the Galapagos' Islands just off South America.

As a result of his time on the Galapagos' Islands Darwin developed his theory of evolution. He was not the only person who developed a theory of evolution.

1. Geospiza magnirostris.
3. Geospiza parvula.
2. Geospiza fortis.
4. Certhidea olivasea.

He noticed that the animals were different in these isolated animals and that the finches had different shaped beaks and that the shape of the beak related to what they ate. The similarities and differences between the species was what he used as evidence for his theory of evolution.

According to his theory of evolution all species were developed from simple life forms more than three billion years ago.

Natural Selection

Darwin developed the theory of natural selection.

Natural selection states that:

1. There is always variation, within a species.
2. Organisms produce more young than will survive.
3. Those best adapted to the environment will survive - survival of the fittest.
4. Those that survive pass on their genes to their offspring.
5. Over many generations, the characteristics of the organism change - or a new species could be formed.

One example of Natural selection is the Peppered Moth.

Before the Industrial Revolution most Peppered moths were pale coloured, these were camouflaged on the lichen on the trees and so were difficult for birds to see.

After the industrial revolution, the buildings and trees became covered in black soot. Now the pale-coloured moth was easy to see, but the dark moth was camouflaged. Now the dark coloured moth was difficult to see, therefore less likely to get eaten and therefore able to breed more.

Once the industrial revolution was over, the pollution was cleaned up and the pale coloured moth returned.

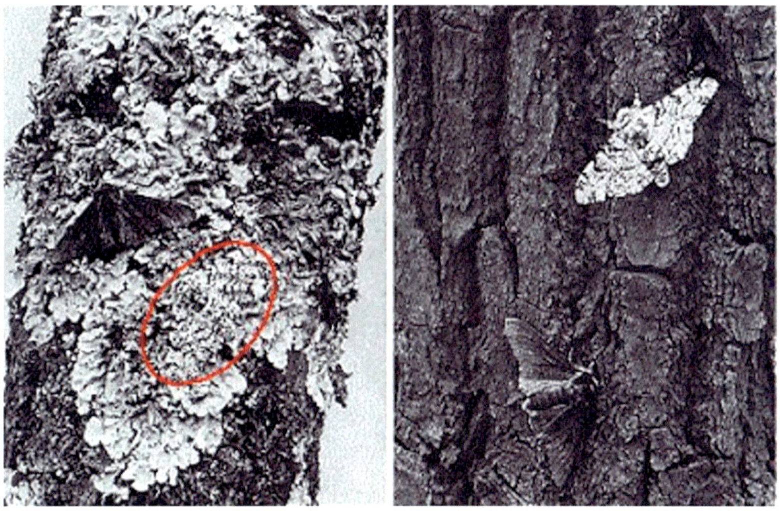

So, somewhere polluted, the dark moth will be more common because:

- The dark moth is camouflaged
- Less dark moths are eaten
- More dark moths survive and breed

Somewhere with little pollution, the pale moth will be more common.

Artificial Selection

For thousands of years people have been breeding plants and animals to produce organisms with desired characteristics. For example, cows have been produced that have lots of milk, wheat has been produced with a large head with lots of seeds. Plants can be produced that resist disease. Artificial selection is also called Selective Breeding.

The process of artificial selection is:

- Choose the parents with the most favourable characteristics.
- Mate.
- Choose the offspring with the most favourable characteristics.
- Mate with other animal with the required favourable characteristics.
- Repeat over many generations.

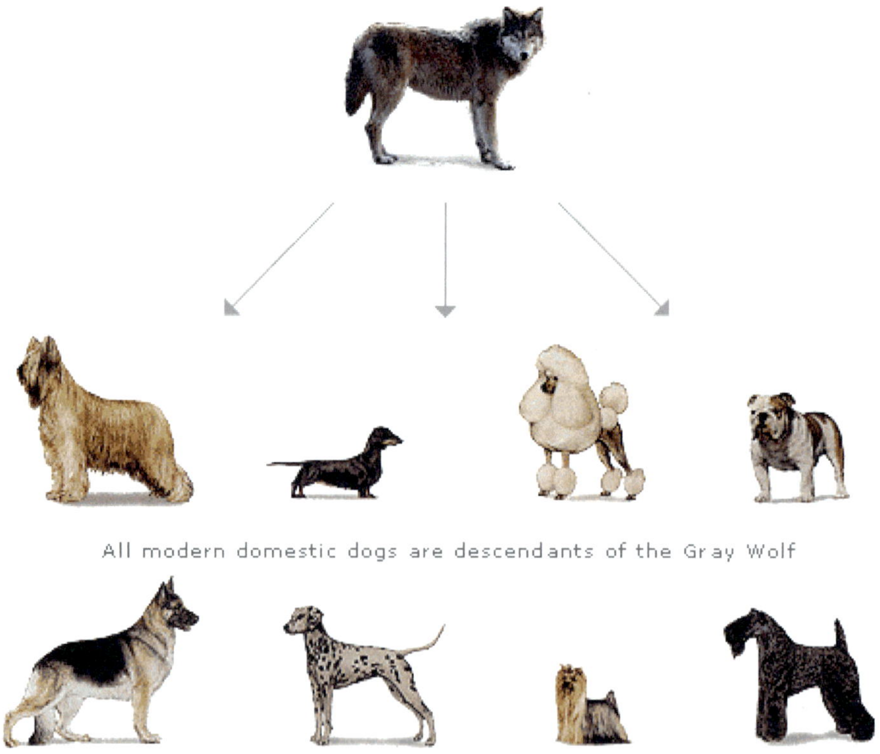

2.3.2 Reproduction

At the end of this topic you should know:

- There are two types of reproduction:
  1. Sexual reproduction, which involves the joining of male and female sex cells. There is a mixing of genetic information, which leads to variety in the offspring
  2. Asexual reproduction, where only one individual is needed as a parent. There is no mixing of genetic information, which leads to identical offspring (clones).

There are two types of reproduction: Asexual and sexual.

Asexual reproduction involves:

- Only one parent
- Offspring are identical (clones) to the parent and each other

Sexual reproduction involves:

- A male and a female sex cell coming together. In animals the male sex cell is called the sperm and the female sex cell is called the ovum or egg. The joining of the sex cells is called fertilisation.
- The genetic material of the parents is mixed. This is what causes the differences between the offspring and their parents.
- Results in variation – that is the parents and offspring are different to each other.

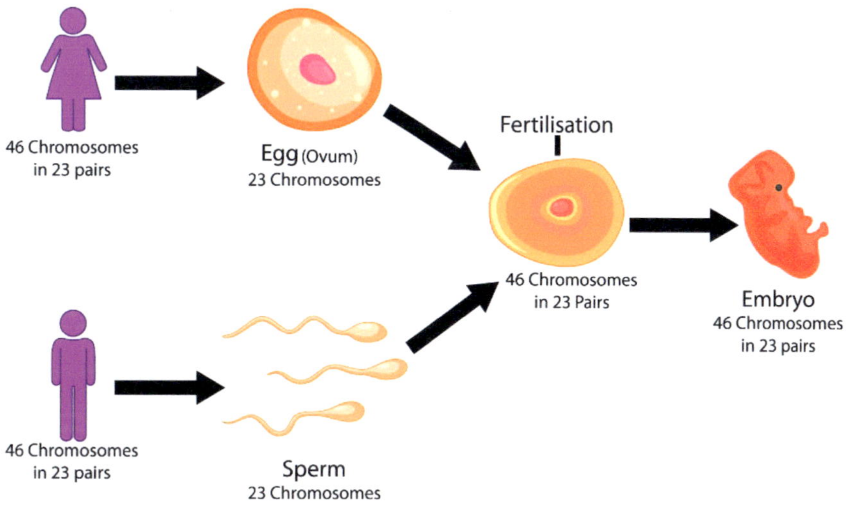

Identical offspring are called clones.

2.3.3 DNA

At the end of this topic you should know:

- The genetic material in the nucleus of a cell is made of a chemical called DNA, which is contained in structures called chromosomes.
- A cell consists of a nucleus that controls the actions of the cell, and cytoplasm.
- Chromosomes carry genes that control the characteristics of the body.
- Humans have 23 pairs of chromosomes. Only one pair carries the genes that determine sex: females have the same sex chromosomes (XX); in males the chromosomes are different (XY).

A part of the cell is called the nucleus. The nucleus contains the DNA, which controls the actions of the cell and makes us who we are.

DNA is a chemical. A molecule of DNA is called a chromosome. Each chromosome consists of lots of genes. Each gene codes for a protein. This controls the characteristics of the body.

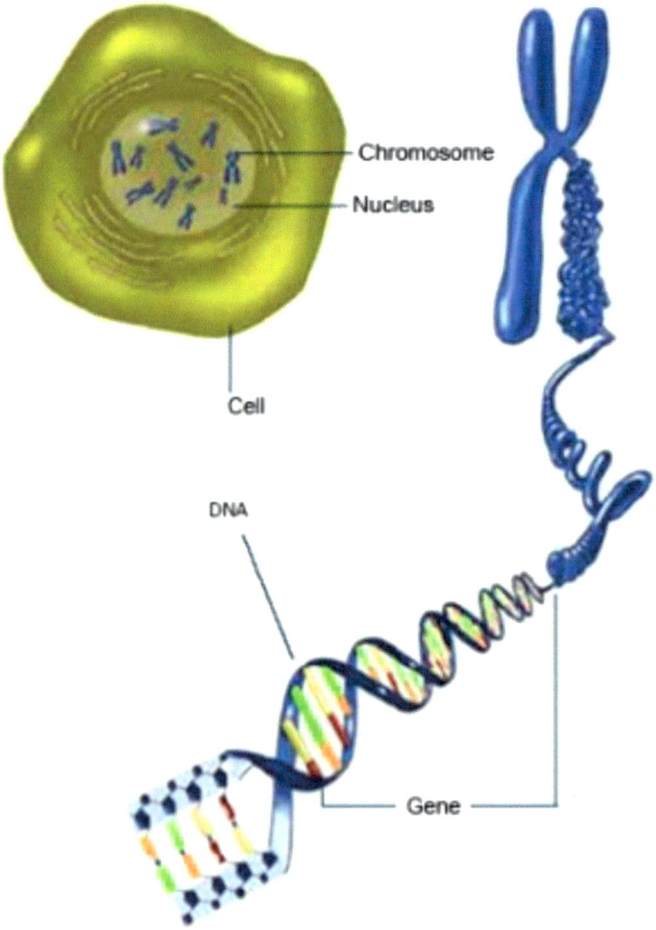

A way to think of this, is to compare it to a string of plastic beads. The plastic is the DNA, each bead represents a gene and the whole string of beads is a chromosome.

Humans have 23 pairs of chromosomes.

They are numbered one to 22. The twenty third pair is called the sex chromosomes. The sex chromosomes determine if you are a boy or a girl.

The sex chromosomes are called X and Y. Females have two X chromosomes. Males have one X and one Y chromosome. The Y chromosome is much smaller than the X chromosome.

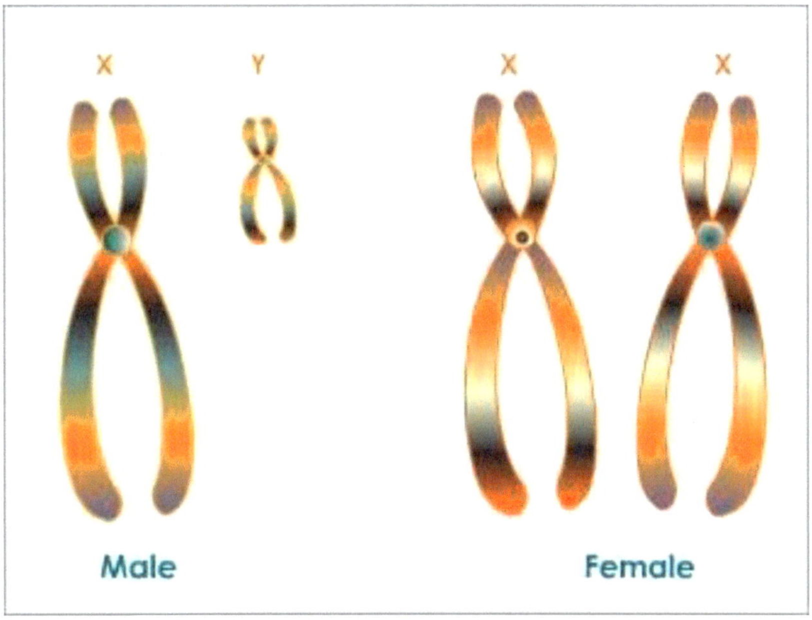

2.3.4 Genetic Engineering

At the end of this topic you should know:

- In genetic engineering, genes from chromosomes of humans and other organisms can be 'cut out' and transferred to the cells of other organisms.
- Be aware of the potential benefits and risks of genetic engineering.

Genes from one organism can be cut out and put into the cells of another organism. This is called genetic engineering.

Genetic engineering often involves bacteria as they can take in DNA easily and they reproduce quickly.

One example of the use of Genetic Engineering is the production of the hormone, insulin.

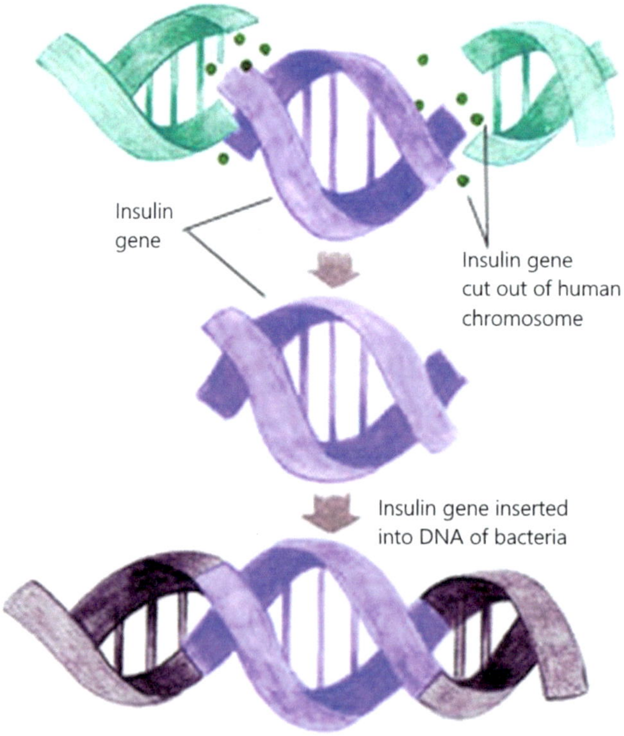

To produce human insulin:

- The insulin gene is cut out of the chromosome.
- The insulin gene is transferred to a bacterium and put into the bacterial DNA.
- The genes work the same as they did in the human cell and produce insulin.

This same process can be used for other genes too.

There are many advantages and disadvantages of using genetic engineering.

Advantages	Disadvantages
Can produce human substances such as hormones. It only takes one generation, compared to selective breeding that takes lots. Can produce organisms with a specific desired characteristic, such as resistance to disease. Can transfer characteristics across species and even kingdoms.	There is a concern that genetically engineered crops could breed with wild plants. Could reduce the natural variation. Long term effects are not known. Some people think that we should not change nature.

Chemistry Component 3 Elements, Mixtures and Compounds

3.1 Atoms, Elements and Compounds

3.1.1 Matter and atoms

At the end of this topic you should know:

- All substances are made of atoms.
- An atom is the smallest part of an element that can exist.
- A substance that is made of only one sort of atom is called an element.
- There are about 100 different elements.
- Elements are shown in the Periodic table.
- Metals are towards the left and the bottom of the Periodic table and non-metals towards the right and the top of the Periodic table.
- Over 2/3 of the elements in the periodic table are metals.
- Elements in the same group of the Periodic table have similar chemical properties.
- Elements in group one are very reactive metals.
- Elements in group 7 are reactive non-metals.

Matter is anything that has mass and takes up space.

All matter is made up of tiny particles called atoms, which are constantly moving.

Atoms are very, very small. So small that there are millions of atoms in just one gram of carbon. While we cannot see atoms, because they are too small, they are often drawn as small balls.

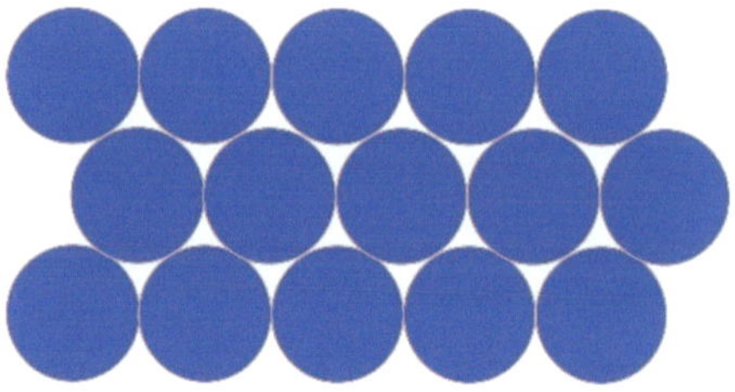

This is a model of a solid. The atoms are represented by circles that are touching and are all vibrating (shaking).

An element is made up of only one type of atom. The atom being the smallest part of an element that can exist . There are about 100 elements.

Compounds have more than one type of atom joined together.

A molecule is two or more atoms joined together. The atoms can be the same, making it a molecule of an element, or different, making it a molecule of a compound.

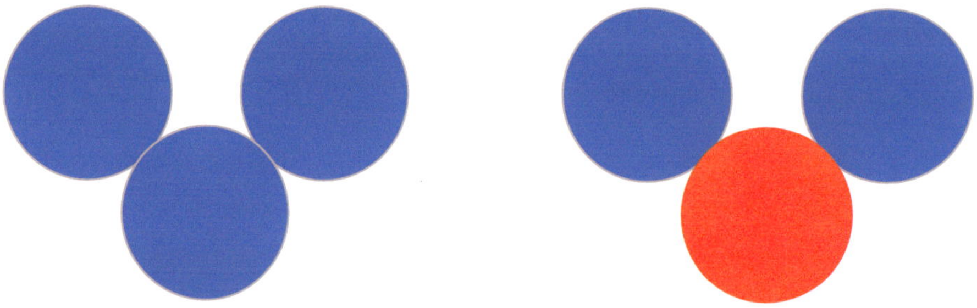

A molecule of an element
(Only one type of atom)

A molecule of a compound
(More than one type of atom)

Metals and Non-metals

Elements can be either metals or non-metals.

Metals are grey, solids. All metals except mercury are solids at room temperature. They look shiny when cut. Metals can be hammered into a sheet or drawn out to make a wire. Metals can conduct electricity and heat. That means when one end gets hot, the rest of the metal will get hot too. So, if you put a spoon in a hot cup of tea, the heat will travel up the handle of the spoon, even though it's not in the tea.

Non-metals are solids, liquids and gases. They are not shiny but dull. They cannot be hammered into a sheet but will crumble if hit too hard, we say that they are brittle. Non-metals do not conduct electricity or heat.

The Periodic Table

The different elements are shown in the Periodic Table. The Periodic Table not only lists the elements but tells Scientists about their properties.

The metals are found on the left and bottom of the periodic table while non-metals are found on the right of the periodic table.

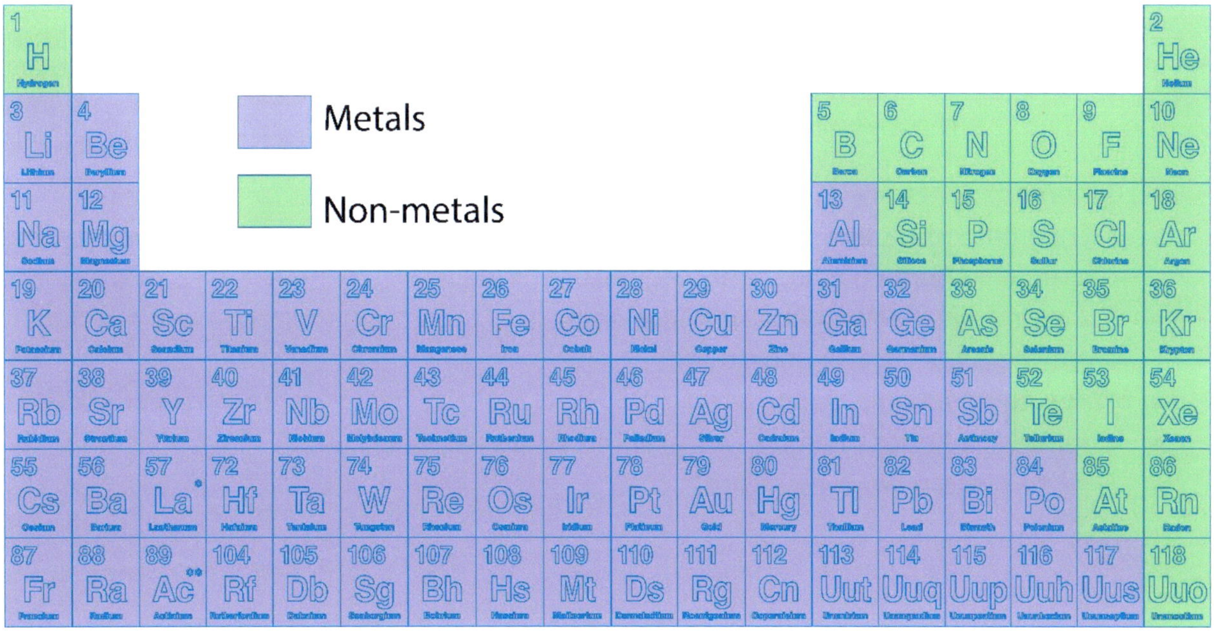

In the periodic table the columns are called groups and the rows are called periods. Elements in the same group have similar chemical properties to each other.

- Group 1 (the first column of the Periodic Table) contains metals that are very reactive. They will even burst into flame when they are put in water. While they are metals, they are soft and able to be cut with a knife. Examples include Sodium and Potassium.
- Group 7 (the second last column of the Periodic Table) contains reactive non-metals. Examples include fluorine and chlorine.

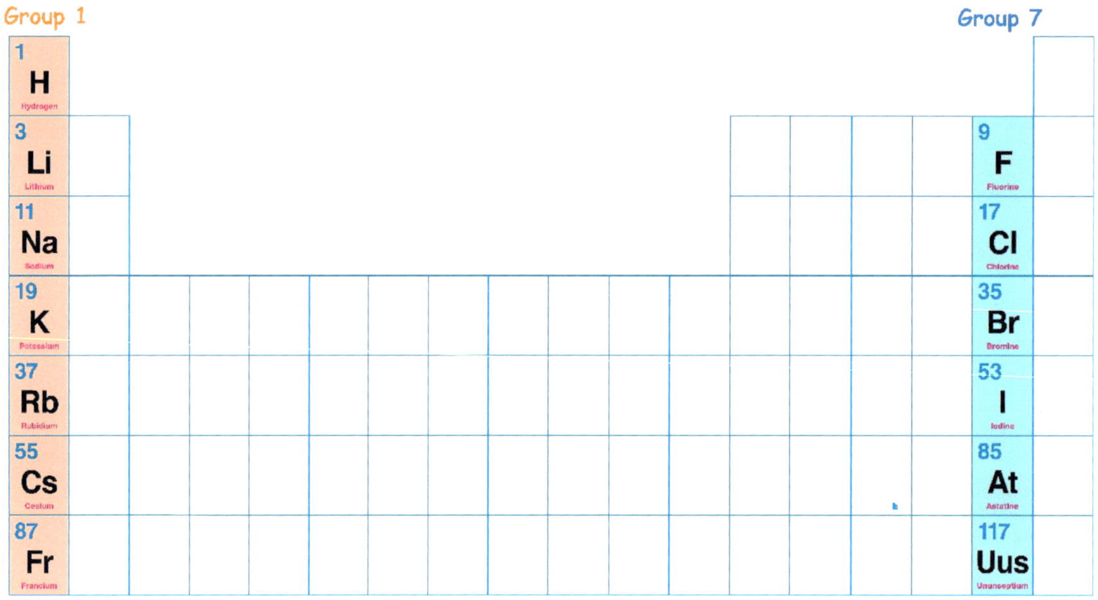

3.1.2 Matter and atoms

At the end of this topic you should know:

- When elements react, their atoms join with other atoms to form compounds.
- Some compounds are made from metals combined with non-metals, for example sodium chloride and magnesium oxide.
- For many simple compounds, the elements in a compound can be worked out from the name of the compound.
- Some compounds are made from only non-metals, for example carbon dioxide.
- Chemical reactions can be represented by word equations.
- How to write a word equation.

Compounds

When atoms of different elements join together, compounds are formed.

Some compounds are made from a metal and a non-metal. For example, sodium chloride is made when sodium and chlorine react together. Magnesium oxide is formed when magnesium and oxygen react together.

Sodium chloride, contains sodium and chlorine, joined together.

Some compounds are made from two non-metals reacting together. For example, carbon dioxide is made from carbon and oxygen.

Naming Compounds

Simple compounds are made from the elements that make them up.

When a compound is made of a metal and a non-metal, then:

- The metal is written first and the name of the metal stays the same.
- The end of the name of the non-metal changes to -ide.

For example, in sodium chloride. The metal is sodium so that is written first. The non-metal is chlorine. The end is changed to -ide making chloride. These two parts are put together to make sodium chloride.

In the same way, a compound consisting of magnesium and oxygen would be magnesium oxide.

Some compounds are made from more than two elements. If a compound contains a metal, a non-metal and oxygen, then:

- The metal is written first and the name of the metal stays the same.
- The end of the non-metal (that's not oxygen) changes to -ate.

For example, in copper sulphate; copper is the metal so is written first. The non-metal is sulphur. The ending is changed to -ate making it sulphate. Oxygen is not named, as the -ate ending tells us that the compound contains oxygen. So, these two parts are put together to make copper sulphate.

Compounds made from two non-metals. The first name is unchanged. The end of the second name is changed to -ide. Sometimes, if the compound contains more than one atom of the element, a pre-fix is added.

The prefixes are:

- Mono- 1
- Di – 2
- Tri – 3
- Tetra – 4
- Penta - 5

So, a compound containing one carbon atom and two oxygen atoms will be called carbon dioxide. A compound containing two sulphur atoms and three oxygen atoms will be called disulphur trioxide.

carbon + oxygen → carbon dioxide

Chemical Reaction

When two or more substances react together, a new substance is formed.

In a reaction, there are the same atoms at the beginning of the reaction and at the end of the reaction, they are simply rearranged.

hydrogen + oxygen → water

Sometimes, elements can react together to form compounds. This can be seen in both of the above reactions, where carbon and oxygen combine to form carbon dioxide and also where hydrogen and oxygen combine to form water.

These reactions can be written as word equations. The substances we start with are written on the left. If there is more than one substance, we put a "+" sign between the substances. The products, that is the substances we make, are written on the right. Between what we start with and the products we draw an arrow.

For example: If we combine nitrogen and hydrogen to form ammonia, the word equation would be written as:

nitrogen + hydrogen → ammonia

3.2 How Structure Affects Properties

3.2.1 States of Matter

At the end of this topic you should know:

- The three states of matter are solid, liquid and gas.
- Melting and freezing take place at the melting point, boiling and condensing take place at the boiling point.
- The three states of matter can be represented by a simple model. In this model, particles are represented by small solid spheres.
- When a solid melts to become a liquid, the particles are able to move about but stay close together.
- When a liquid boils and becomes a gas, the particles separate and move about rapidly.
- Substances with high melting points have strong forces that hold their particles together.
- Substances with low boiling points have weak forces between their particles.

The states of matter

All matter can exist in three states: solids, liquids and gases.

Solids

The particles in a solid are all touching. They are arranged in a regular pattern; they are constantly vibrating (shaking) but cannot move out of position.

Liquids

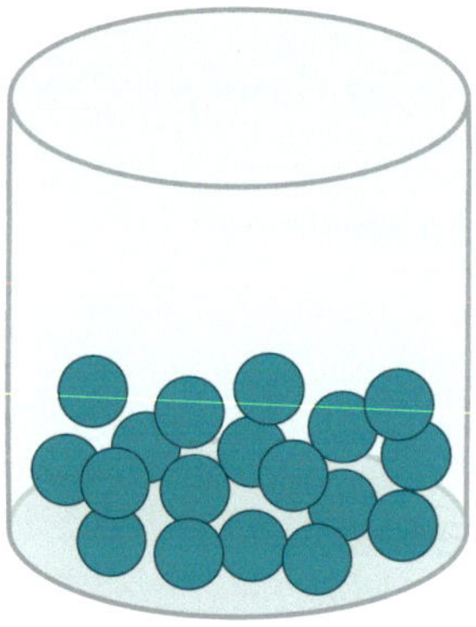

The particles in a liquid are touching but they are not arranged in a regular pattern. They can roll over each other.

Gases

The particles in a gas have lots of space between them. They move quickly in all directions.

Change of State

When a substance is heated or cooled its state can change.

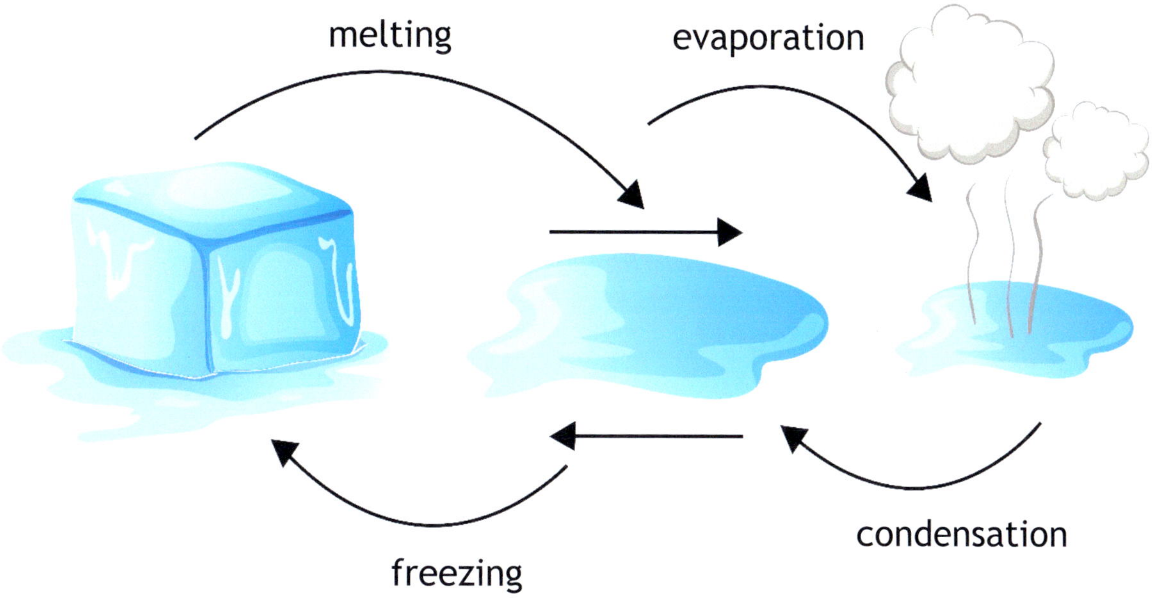

When a substance is heated (energy is added):

- A solid becomes a liquid - this is called melting.
- A liquid becomes gas - this is called evaporation

When a substance is cooled (energy is removed):

- A gas becomes a liquid - this is called condensation#
- A liquid becomes a solid - this is called freezing or solidification.

Substances melt and freeze at the same temperature. This is called the melting point. The melting point of water is 0°C.

Substance boil and condense at the same temperature. This is called the boiling point. The boiling point of water is 100°C.

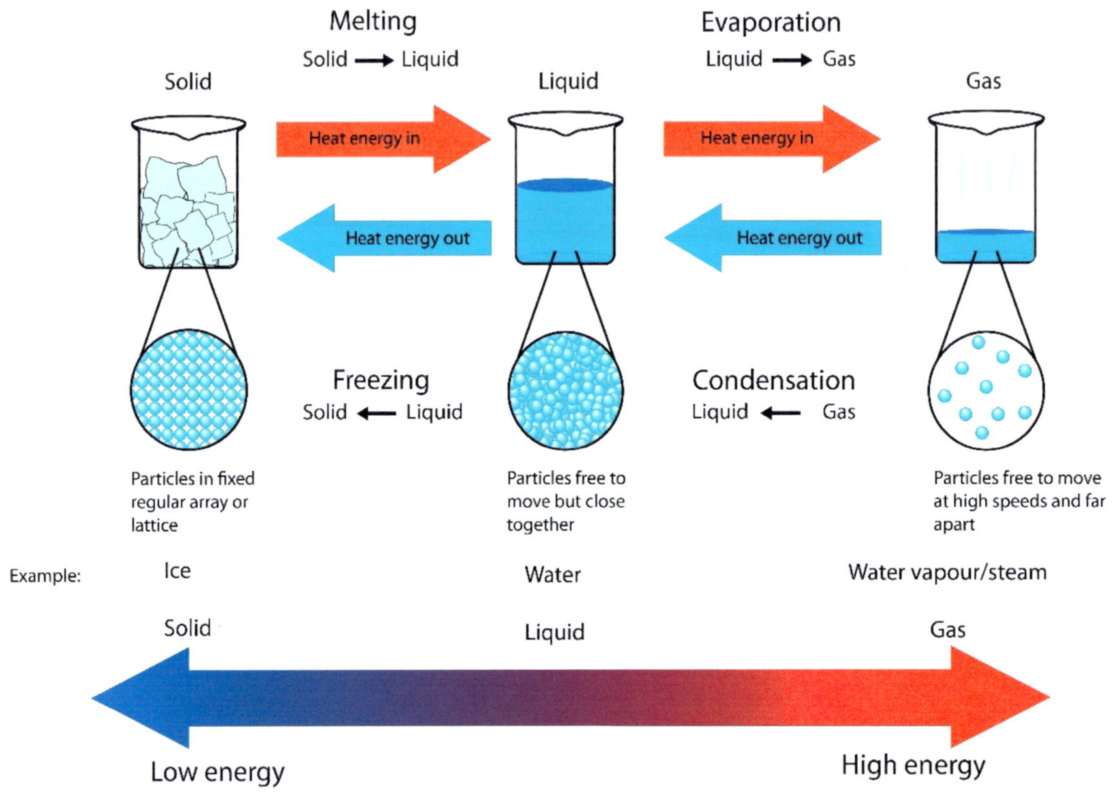

Changes of state are a physical change. No new substances are formed. They are reversible.

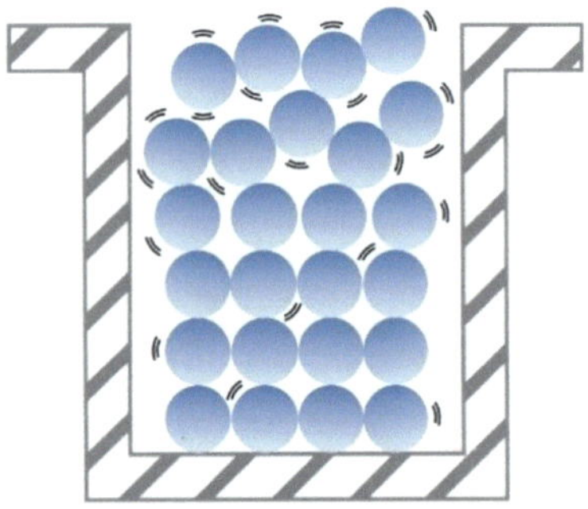

Melting

Melting happens when a solid is heated. The particles move faster and more strongly. This causes the bonds between the particles to break. When the substance melts, the particles break away from their fixed positions and move around.

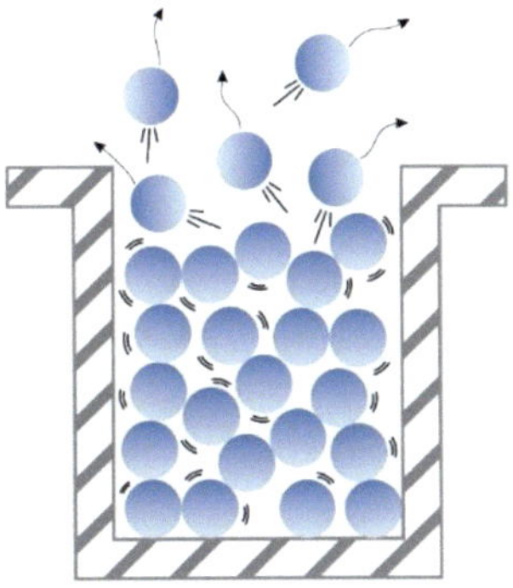
Evaporation

Evaporation occurs when the particles in a liquid gain enough energy to break away from the liquid and become a gas.

When substances are cooled, in condensation and freezing, the particles move closer together and stronger forces are formed between the particles.

The stronger the forces between the particles in a substance the higher the melting point will be.

- Substances with strong forces between their particles have high melting and boiling points.
- Substances with weak forces between their particles have low melting and boiling points.

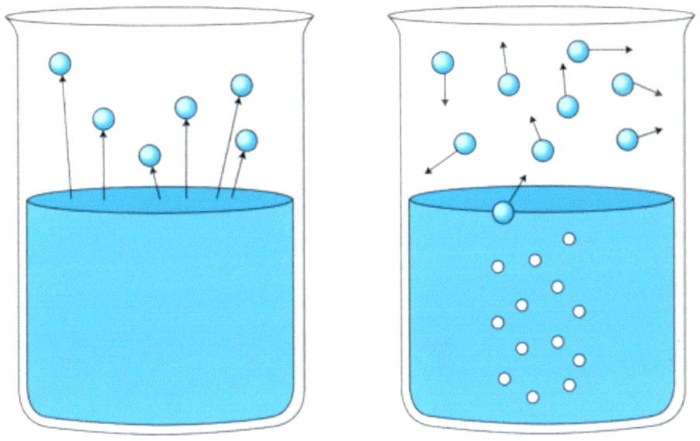
Evaporation Boiling

Evaporation and boiling both refer to liquid becoming gas. However, they are different because:

- Evaporation happens at any temperature – not just the boiling point.
- Evaporation only happens at the surface of the liquid, unlike boiling which happens throughout the liquid.

3.2.2 Carbon

At the end of this topic you should know:

- Diamond and graphite are forms of the element carbon with different properties because of their different structures.
- Diamond is hard because the carbon atoms are joined together in a giant three-dimensional structure.
- Graphite is slippery because the carbon atoms are joined together in layers that can slide over each other.

Carbon is a non-metal element. Carbon can join with other carbon atoms to make very large 3D structures. Carbon is found in a number of forms. Two of those forms are diamond and graphite. Diamond and graphite have very different properties because the carbon atoms are joined together differently resulting in different structures.

Diamond

In diamond, each carbon atom is joined to four other carbon atoms, creating a giant network. The bonds between the atoms are very strong, result in a very strong, rigid structure. Diamond is a very hard, strong material because of the strong bonds. Diamonds are shiny crystals that do not conduct electricity.

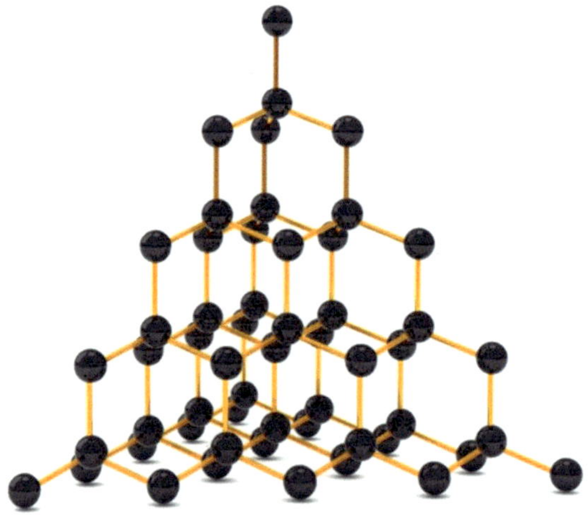

Diamonds are used for cutting, such as in glass cutters and rock drilling equipment. They are also used in jewellery.

Graphite

In graphite, each carbon atom is joined to three other carbon atoms. The atoms are arranged in hexagons which form layers. Since they are joined to three atoms, they have an extra electron, called a delocalised electron. These electrons can move allowing graphite to conduct electricity.

While the bonds within the layers are very strong there are no strong bonds between the layers. This means that the layers can slide over each other. Graphite is soft and slippery.

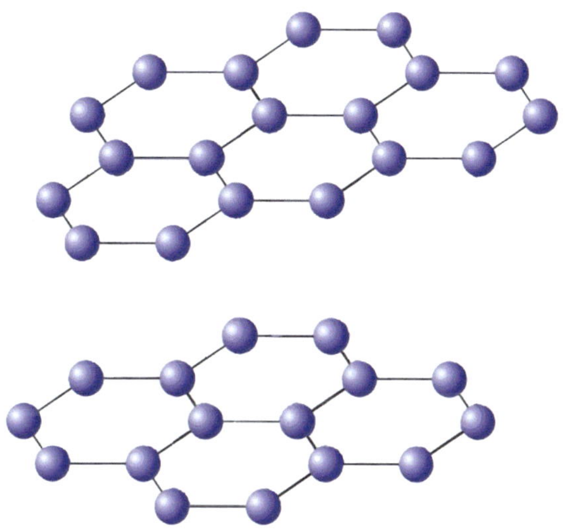

Graphite is used in lubricants to help surfaces slide over each other and also in some pencils.

3.3 Separating Mixtures

3.3.1 Mixtures

At the end of this topic you should know:

- A mixture contains two or more substances not chemically combined.
- Mixtures can be separated by processes such as filtration, distillation, crystallisation and chromatography.
- Paper chromatography can be used to separate mixtures and can give information to help identify substances.
- In paper chromatography a solvent moves through the paper carrying different compounds different distances.

Mixtures

In chemistry a pure substance is one that contains only one substance, either an element or a compound. Therefore, pure water contains only water molecules, pure oxygen contains only oxygen molecules and pure gold contains only gold atoms.

A pure substance always has a fixed composition. For example, water always contains two hydrogen atoms for every oxygen atom. Sodium chloride always contains one sodium atom for every chorine atom.

Water molecules

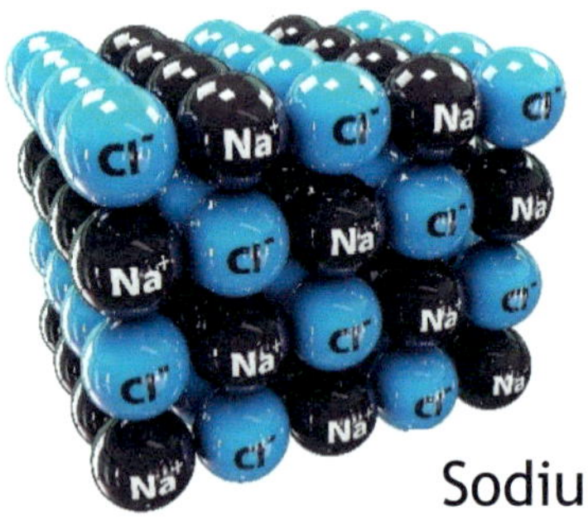

Sodium chloride

A pure substance has a fixed melting and boiling point. For example, water always boils at 100°C and melts at 0°C.

A mixture contains two or more substances not chemically combined. Those substances can be elements, compounds or a mixture of both.

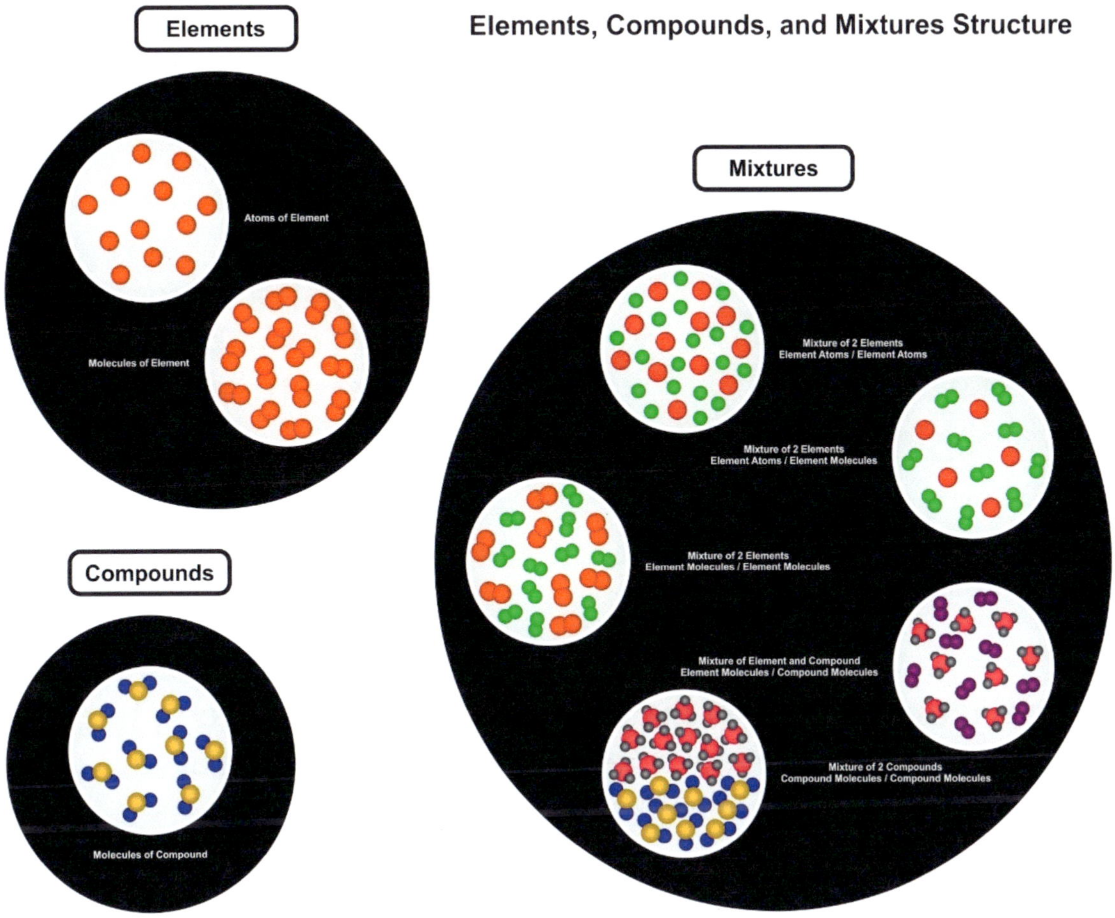

Unlike, compounds mixtures do not have a fixed composition. For example, salt water can contain different concentrations of salt and you can have a weak cup of coffee or a strong one.

In a mixture the substances have the same properties as they do separately. This is different to a compound where the compound can have vastly different properties to the elements that make it up.

Let's look at iron and sulfur as an example. Iron is a grey, magnetic metal, while sulfur is a yellow solid.

Image of an iron sulfur mixture. From Creative Commons. Attribution on back page.

The iron and the sulfur keep their own properties. Therefore, the iron is still magnetic and can be separated from the sulfur by a magnet.

Image of an iron sulfur mixture. From Creative Commons. Attribution on back page.

However, when heat is added and the iron and sulfur react, they form iron sulphide, which does not have the properties of iron or sulfur and is not magnetic.

The melting point and boiling point of a mixture depends on the concentration of each of the substances in the mixture. Mixtures normally melt and boil over a range of temperatures.

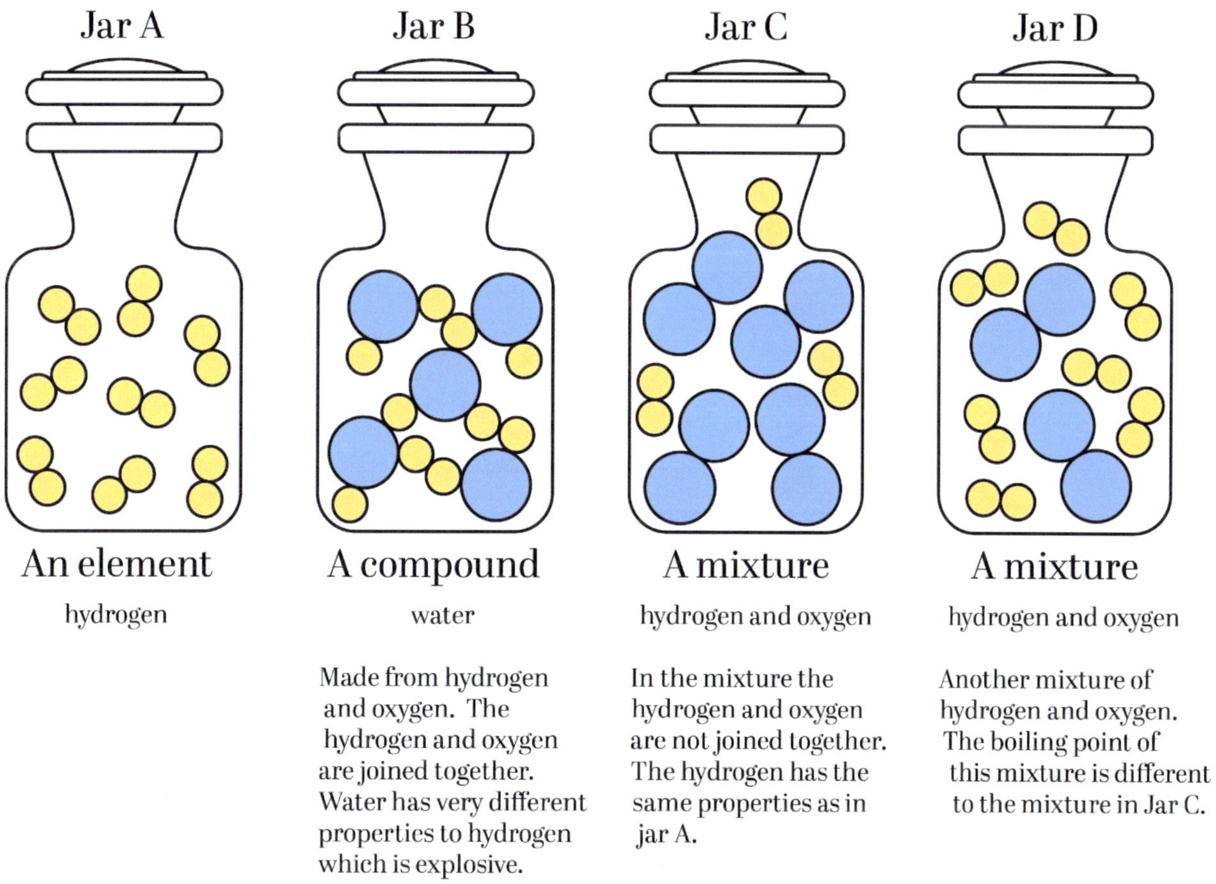

Solutions

A solution is one type of mixture. In a solution the particles that are dissolved are so small that the solution is always clear, and the particles do not settle out over time. Some substances do not dissolve. They are called insoluble. They still form a mixture but not a solution.

Solution components

Solute	-	the solid being dissolved.
Solvent	-	the liquid it is being dissolved in.
Soluble	-	means it will dissolve e.g. salt is soluble in water.
Insoluble	-	means that it doesn't dissolve.

Making a salt water solution by dissolving table salt (NaCl) in water.

The salt is the solute and the water the solvent.

From Creative Commons. Attribution on back page.

Separating Mixtures

Mixtures can be separated. The method of separation depends on the mixture that is being separated.

Method	Diagram	When used
Filtering	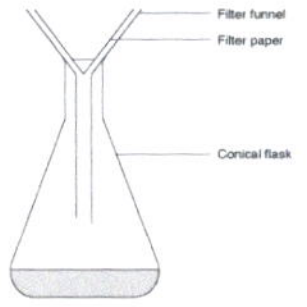	A liquid and an insoluble solid.
Sieving	***	Two solids of very different size
Crystallisation	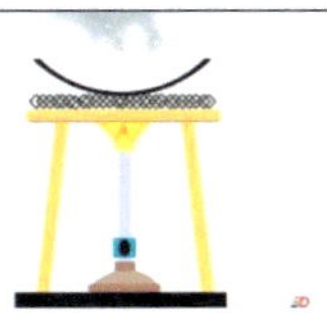	A solution where only the solute is wanted.
Distillation	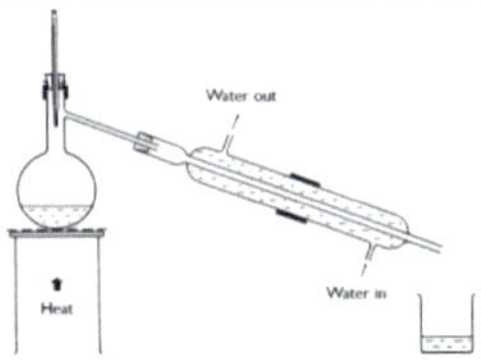	A solution where the solvent is wanted; or two liquids with different boiling points.
Chromatography	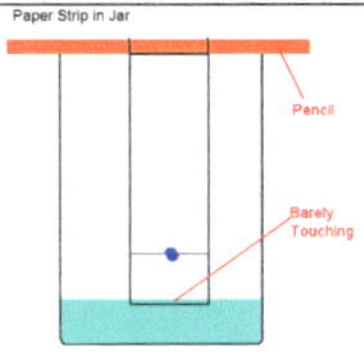	Anything with a different size or charge. Name comes from the separation of colours.

In filtration a filter paper is put into a filter funnel. The liquid can pass through the filter paper, but the solid particles, such as sand are too big and stay in the filter paper.

So, in this example, the water and dissolved coffee can go through the filter paper, but the coffee grounds are too big to fit through the holes in the filter paper and stay in the filter paper.

In distillation the mixture is heated. The substance with the lowest boiling point will evaporate first. When it becomes a gas, it will spread out everywhere including the condenser. The condenser is surrounded by a water jacket which cools the gas. This causes the gas to turn back into a liquid (condense). The liquid then runs down the condenser to the container waiting to collect it.

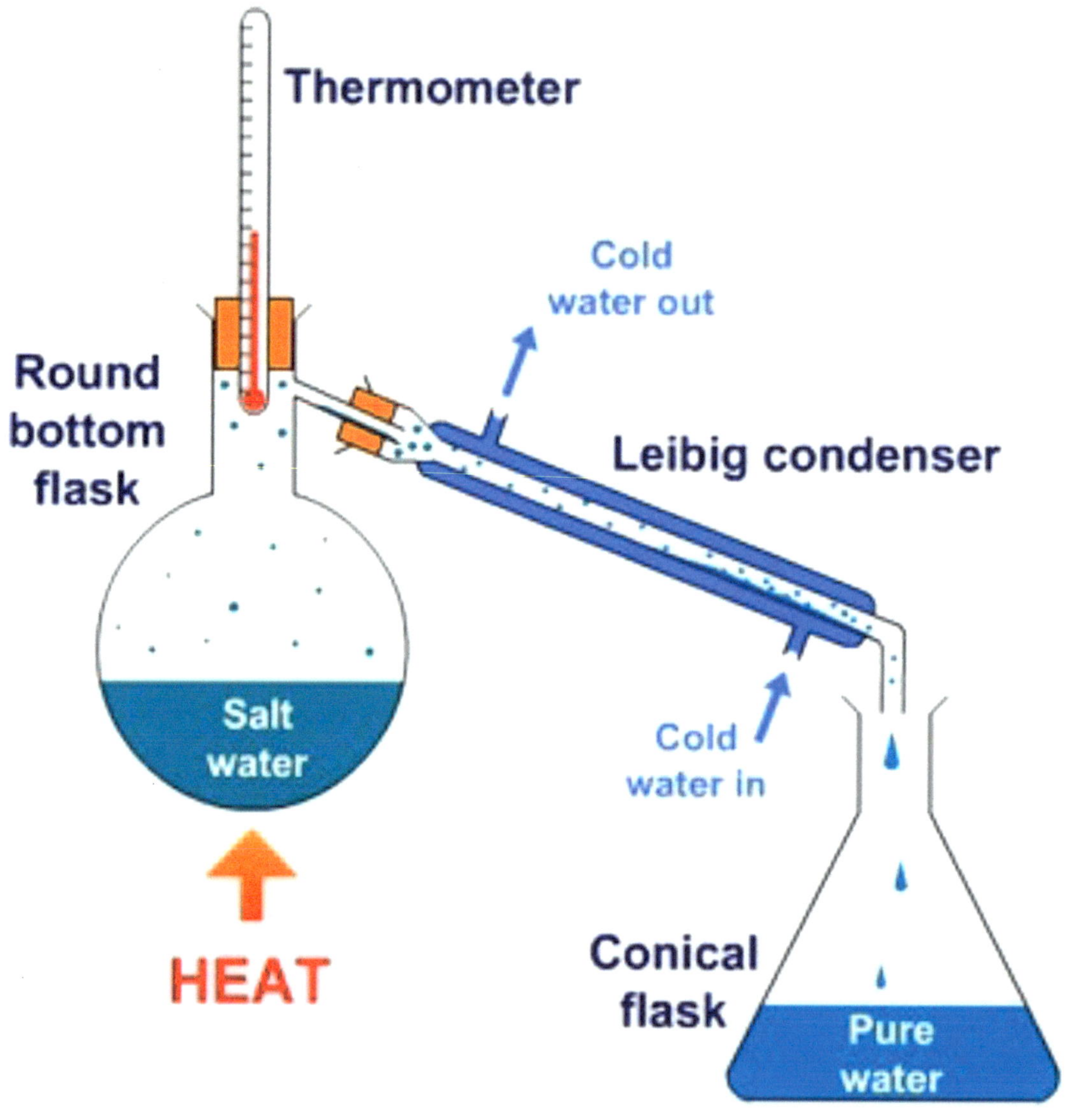

In crystallisation a solution is heated. The solvent is evaporated leaving the solid solute behind.

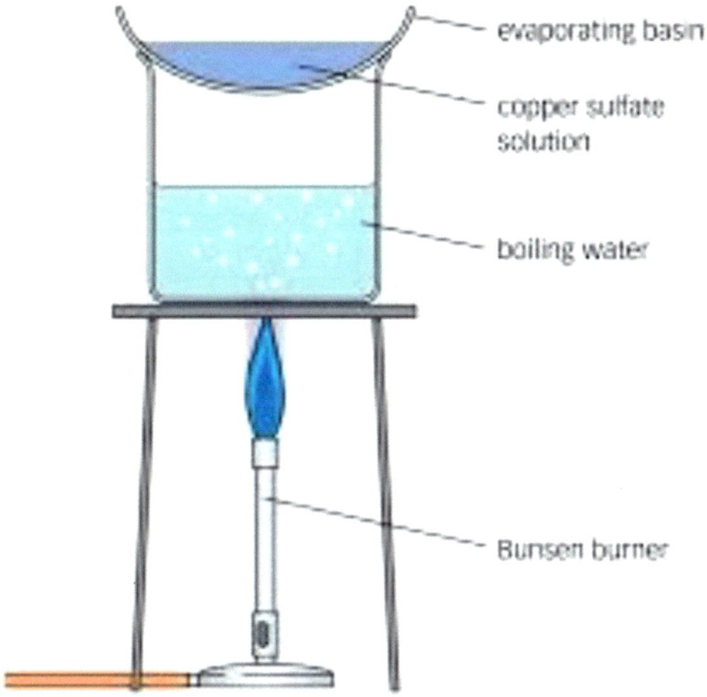

Paper Chromatography

In paper chromatography:

- A line is drawn in pencil about 1 cm from the bottom of the strip of filter paper. It must be in pencil as ink would dissolve in the solvent and run up the paper.
- One or more dots of the solution is placed on the line.
- Put a small amount of the solvent in a beaker. This is normally water or ethanol.
- Put the filter paper in the beaker. The liquid must not be higher than the dots of the solution being separated. If the liquid was higher the dots would simply dissolve in the liquid.
- Leave in one place, without moving it, until the liquid is near the top of the filter paper.
- Remove and have a look at the dots.

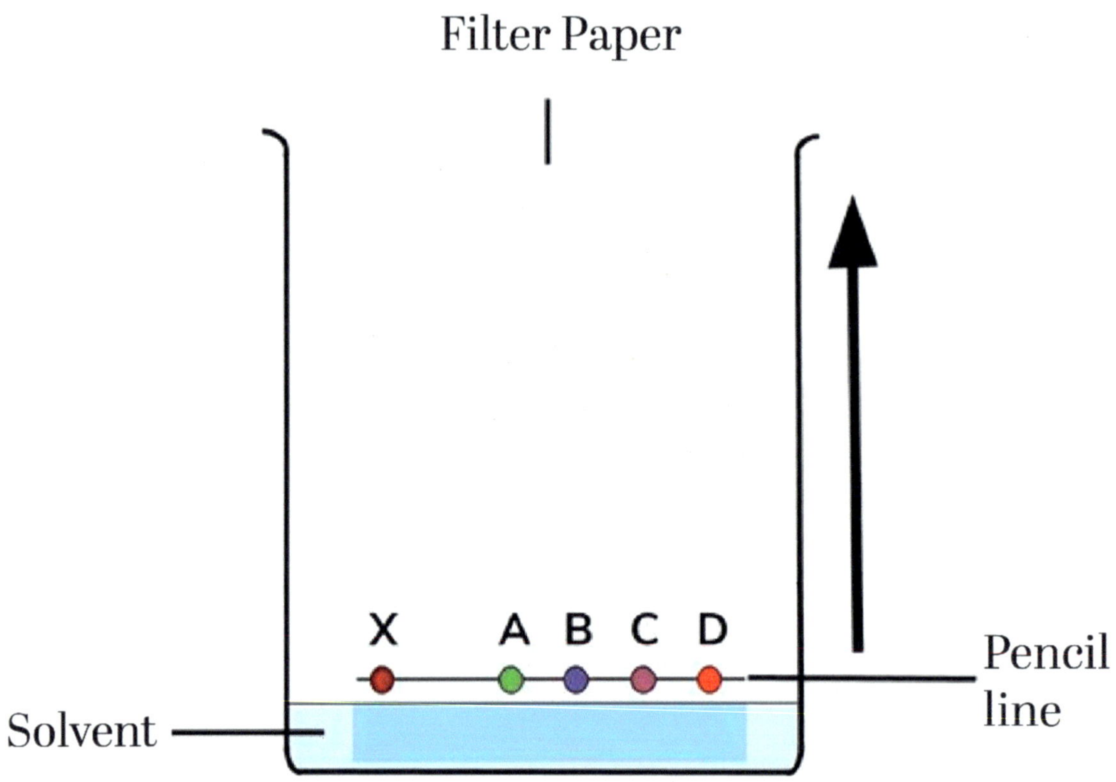

In this example, x is an unknown and will be compared to the known samples.

Different substances will move up the chromatogram by a different amount.

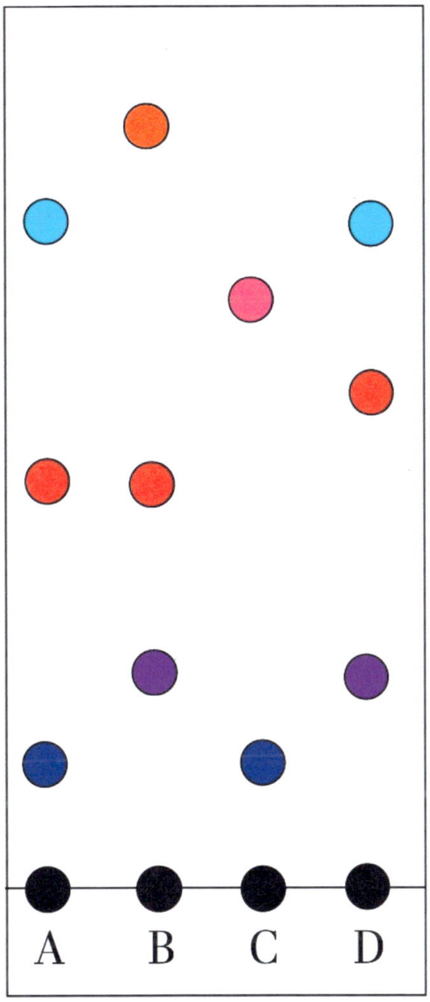

In a chromatogram, the number of spots is the number of substances in the solution. So, in the above chromatogram, Samples A, B and D contain three substances while C only contains two.

Substances that move the same distance are the same. So, the blue colour in samples A and C is the same, as is the purple colour in samples B and D. However, while the red colour in A and B is the same, even though the red in sample D looks the same it is a different substance because it travelled a different distance.

3.4 Metals and Alloys

3.4.1 Extracting, Using and Recycling

At the end of this topic you should know:

- Unreactive metals, such as gold, are found in the earth as the metal itself, but most metals are found as compounds that require chemical reactions to extract the metal.
- Metals less reactive than carbon can be produced by heating the metal compounds in the ore with carbon.
- Ores contain enough metal to make it economic to extract the metal. Large amounts of rock need to be quarried or mined to get metal ores.
- We should recycle metals to save resources and limit environmental impacts.
- There are social, economic and environmental impacts of mining ores and recycling metals.

Reactivity

Different metals have different reactivity. Some metals such as gold, silver and platinum will not react with water, air or other substances. As a result, gold, silver and platinum can be found in the Earth at the metal itself. We refer to these metals as unreactive.

Most metals are reactive and will react with chemicals in the environment such as water and oxygen, to form compounds. Some metals react faster or more violently than others. The order of reactivity is the same regardless of what they are being reacted with. This is called the reactivity series.

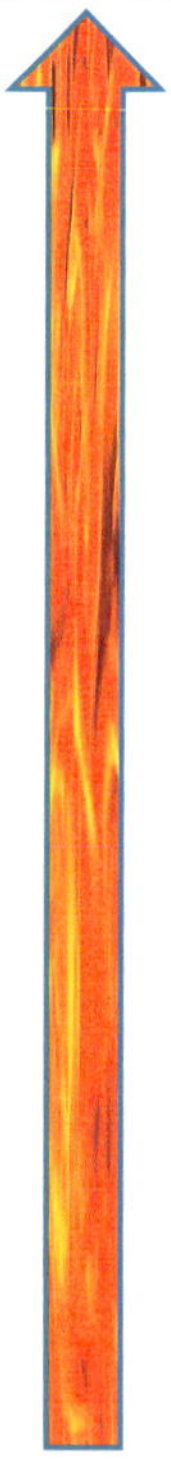

Potassium
Sodium
Calcium
Magnesium
Aluminium
Carbon
Zinc
Iron
Tin
Lead
Hydrogen
Copper
Silver
Gold
Platinum

Metals that are less reactive than carbon can be extracted by heating with carbon.

Metals that are more reactive than carbon use electricity to extract them.

Ores

An ore is a rock that contains enough metal to make it worth extracting. This means that the metal is worth more than the cost of removing the rock from the earth and extracting the metal.

Ores are dug out of the ground from quarries or mines. A quarry is a large deep pit from which rock is cut out. Often a large amount of rock has to be dug out to get just a small amount of ore.

Recycling

There is a limited amount of the different metals in the Earth's crust. It also takes a lot of energy and is very expensive to find, extract and purify metals.

However, metals can be recycled. That means instead of using them once, they can be melted down and used again.

Recycling a metal involves:

- Collecting the metals
- Sorting the metals into different types
- Breaking them up into small pieces
- Removing impurities from the metals
- Using the metals to make new items.

There are many advantages of recycling a metal rather than mining and extracting a metal from its ore. These include:

- Less energy is used. This is important because fossil fuels are also limited and have a negative effect on the environment.
- Fewer mines and quarries. Mines and quarries can damage the environment, cause noise and create traffic.
- Saves the metals in the earth's crust, meaning that the metals in the Earth will last longer.
- It's cheaper.

3.4.2 Properties of Metals

At the end of this topic you should know:

- Metals have giant structures of atoms with strong bonds between the atoms and so most metals have high melting points.
- Metals are good conductors of electricity and thermal energy.
- Copper has properties that make it useful for electrical wiring and plumbing.
- Aluminium is a useful metal because of its low density and resistance to corrosion.

Structure

Properties of metals include:

- Strong
- Solid at room temperature (except mercury)
- Shiny
- Conduct electricity
- Conduct heat
- Can be drawn into a wire
- Can be hammered into a sheet.

These properties make metals very useful. These properties are because of their structure.

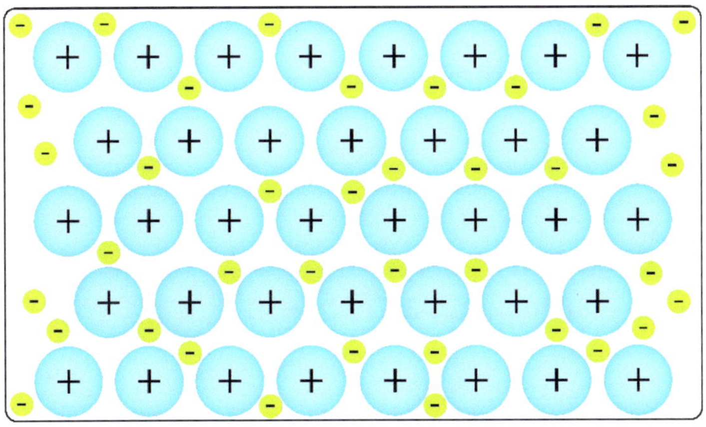

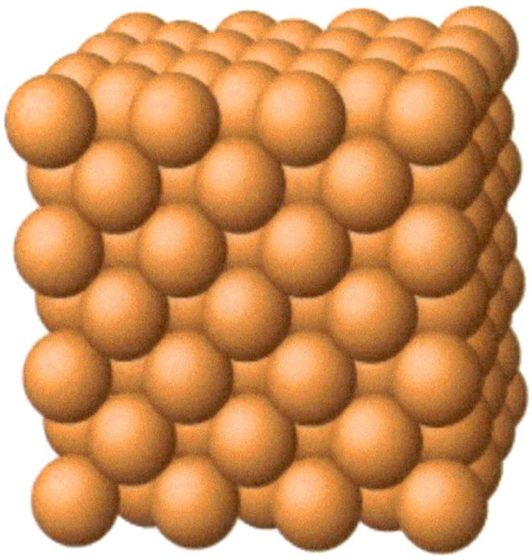

Metals form a large structure with strong bonds holding the atoms together. This results in the metal being strong and having a high melting and boiling point.

In this structure, some of the tiny negative particles in the atoms, called electrons can move through the structure. This allows the metal to conduct electricity and heat.

The atoms of metals form layers. These can slide over each other. This allows metals to be shaped including being drawn into wires and hammered into sheets.

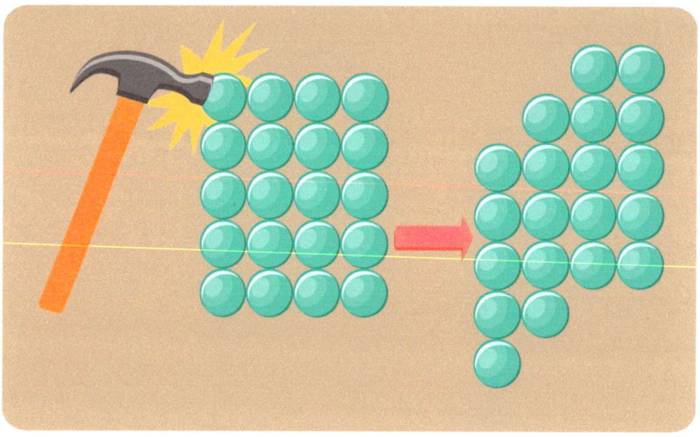

Copper and aluminium

Copper is very good at conducting electricity. It is also very good at being hammered into shape and being drawn into wires. Copper is used for most wiring in houses.

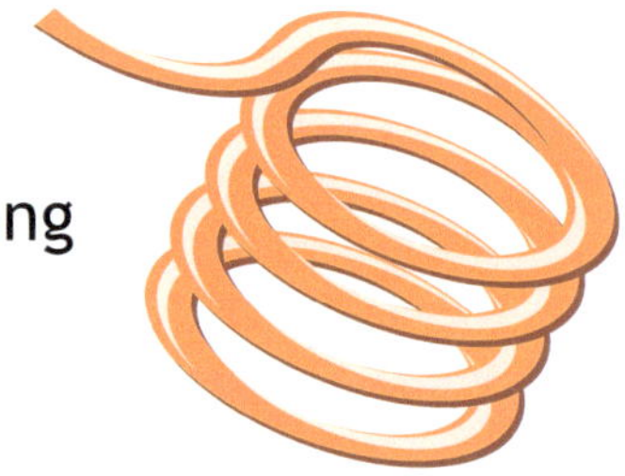

Copper is also very unreactive. It does not react with water. Copper is used for water pipes, especially hot water pipes as it doesn't react with the water.

Aluminium is very useful because it is light and strong. It has a low density, which means it's very light compared to the same amount of other metals. This means it can be used for aeroplanes and it will be able to fly easier and use less fuel.

It also doesn't corrode easily. If a metal corrodes the surface wears away.

Aluminium can be hammered into thin sheets and shaped. Aluminium foil and take-away containers are common in many households.

3.4.3 Alloys

At the end of this topic you should know:

- Most metals in everyday use are alloys.
- Pure iron, gold and aluminium are too soft for many uses and so are mixed with small amounts of other elements to make alloys, which are harder for everyday use.
- Most iron is converted into steels.
- Steels are alloys since they are mixtures of iron with carbon and other metals.

Alloys

Mixtures of metals, or mainly metals are called alloys. Alloys are very useful and most metals that are used in everyday life are alloys.

Some common alloys include:

Alloy	Contains
Bronze	Copper and tin
Brass	Copper and zinc
Steel	Iron and carbon
Stainless steel	Iron, carbon and chromium

A lot of metals are too soft to use on their own. However, alloys are much harder than the metals that make them up. This is because the atoms of the metal that's added has a different size, so disrupts the layers in the structure of the metal.

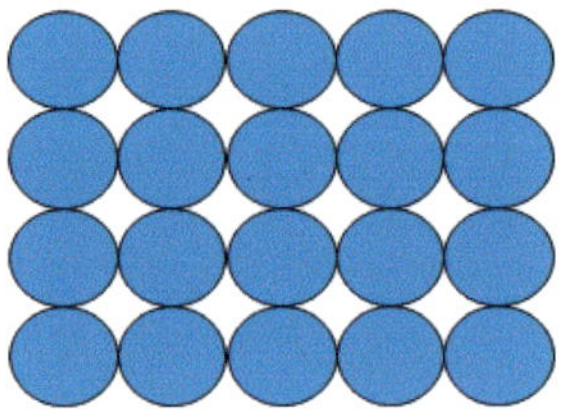

Pure metal

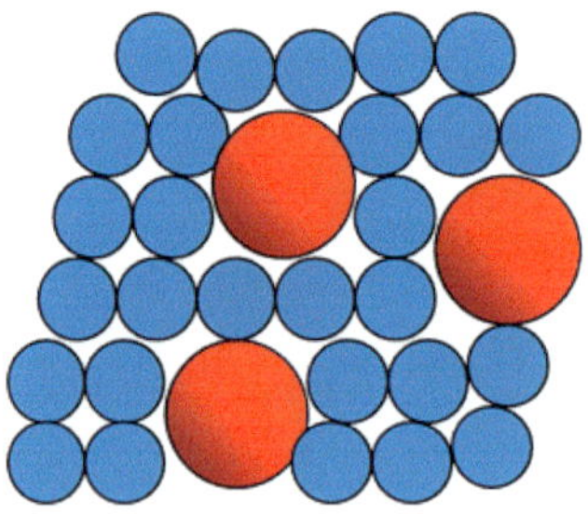

Alloy

Steel

Pure iron is a silvery-white metal that's easy to work and shape. Iron is one of the three magnetic metals. It is too soft and reactive to be very useful.

Steel is an alloy of iron and carbon, and sometimes other elements. Different types of steel have different uses and the elements added to the iron give the steel those properties.

One type of steel is called stainless steel. Stainless steel contains chromium which prevents it from rusting. It normally also includes other elements such as nickel and manganese. However, like all steels it is over 90% iron.

Steel has many uses from reusable drinking straws, to mixing bowls, mixing bowls, car parts, garage doors and bridges.

3.5 Polymers

3.5.1 Polymers

At the end of this topic you should know:

- Polymers such as poly(ethene), poly(propene) polystyrene and PVC are made from small compounds called monomers that join together to form very long chains.
- Polymers are waterproof, resistant to chemicals, and can be moulded, so they have many useful applications as: packaging materials, pipes and containers.
- Many polymers are not biodegradable, so they are not broken down by microbes. This can lead to problems with waste disposal.

Polymers

Polymers are long chains made up of repeating units. These repeating units are called monomers.

Monomers join together to make polymers in a reaction called polymerisation.

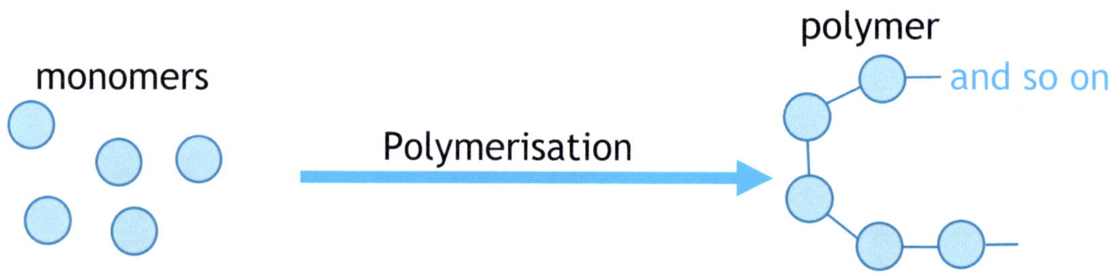

Some examples of polymers are polyethene, polypropene, polystyrene and PVC. Polyethene is made up of lots of ethene molecules joined together.

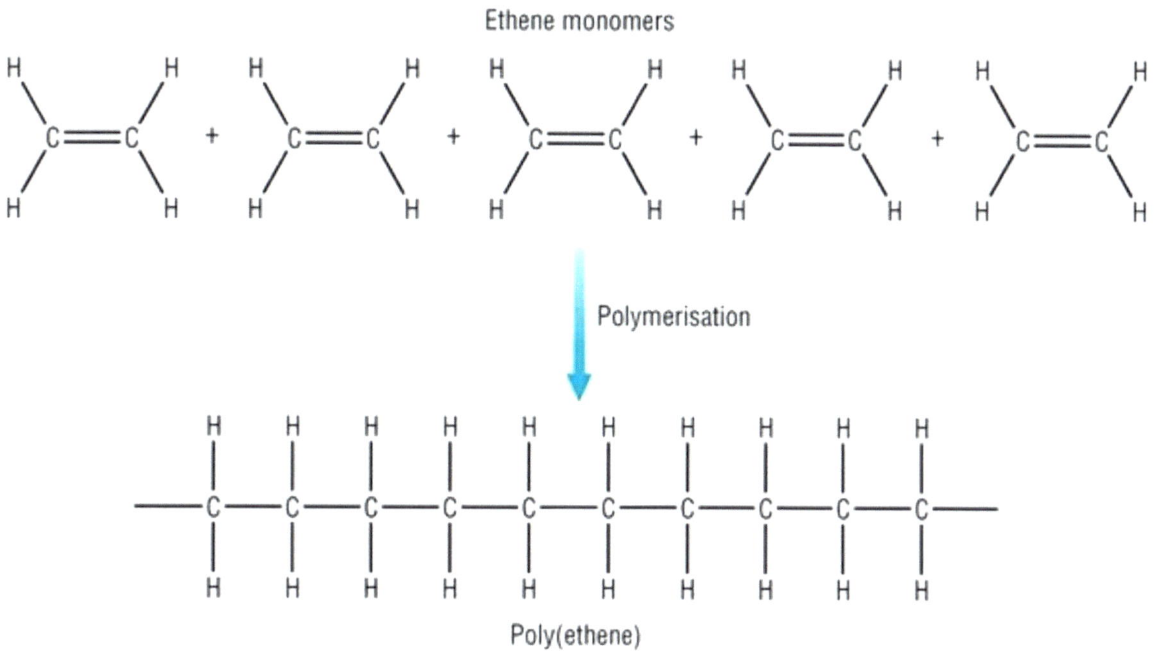

Similarly, polypropene is made from propene and polystyrene from styrene. These substances are called plastics.

This white foam is called polystyrene. It is commonly used to keep things cool or hot for a period of time.

Properties of Polymers

Polymers are waterproof and can be easily moulded into different shapes. They are not very reactive, so they do not corrode or rust like metals do. They are light weight. These properties make them very useful.

Polymers are often used in packaging, pipes and containers.

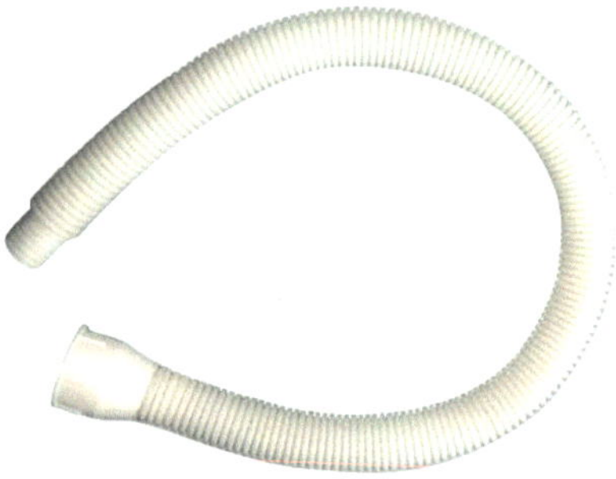

#

The Problem with Polymers

While plastics are very useful, most are not biodegradable. That means that they do not break down over time. However, some can be broken down by microbes and so are biodegradable.

Most polymers end up in landfill.

Polymers can be burnt, but this produces carbon dioxide and other pollution.

This means that plastics that are no longer wanted can last thousands of years and are very difficult to dispose of. This has caused a lot of problems in the environment, particularly in our seas and oceans. A lot of people are trying to put a stop on single use plastics.

Many plastics can be recycled, and this is a lot better option, for the environment, than landfill.

Chemistry Component 4
Chemistry in Our World

4.1 Reactions of Acids

4.1.1 Metals, Acids and Hydrogen

At the end of this topic you should know:

- Acids react with some metals to produce salts and hydrogen.
- Hydrochloric acid produces chlorides and sulfuric acid produces sulfates.
- Be able to complete word equations for these reactions, given the names of the reactants.
- The test for hydrogen uses a burning splint held at the open end of a test tube of the gas. Hydrogen burns rapidly with a pop sound.

Salts

Salts have two parts:

- A metal
- A part from an acid

Different acids form different salts.

- Hydrochloric acid forms chlorides
- Sulfuric acid forms sulfates.

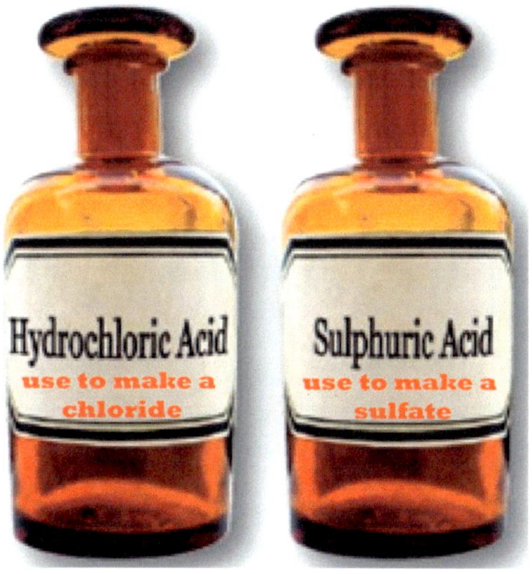

So, work out the name of a salt you need to work out what metal is involved and what acid.

For example, is sodium oxide reacts with sulfuric acid, what salt is formed?

The metal is sodium and the acid is sulfuric acid which produces sulfates, so the salt will be sodium sulfate.

The Burning Splint Test

The test for hydrogen is to put a lit splint into the opening of the test-tube where there is hydrogen gas. A squeaky "pop" sound is heard.

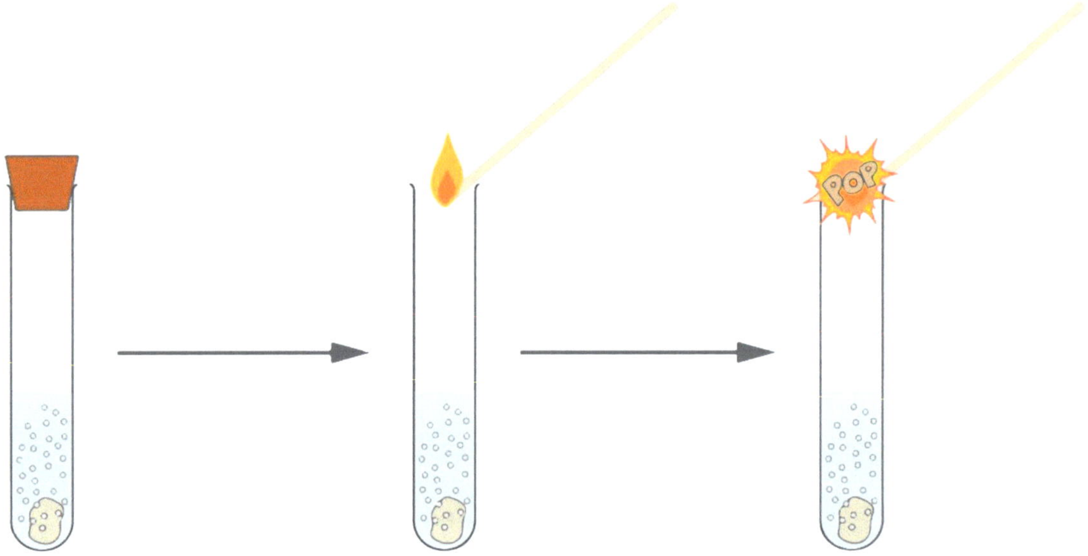

Metals and Acids

Metals react with acids to form salt and hydrogen. The hydrogen will escape into the environment causing the mass of the test-tube or beaker to decrease.

We can write this as the general equation below.

General Equation

metal + acid → salt + hydrogen

For the equation:

- The metal must be the same metal on the left as in the salt
- The salt is worked out from the metal and the acid.
- The hydrogen will be formed as a gas and leave the reaction mixture.

For example, if lithium is added to hydrochloric acid the equation will be:

lithium + hydrochloric acid → lithium chloride + hydrogen

In the next section we will be taking a closer look at acids.

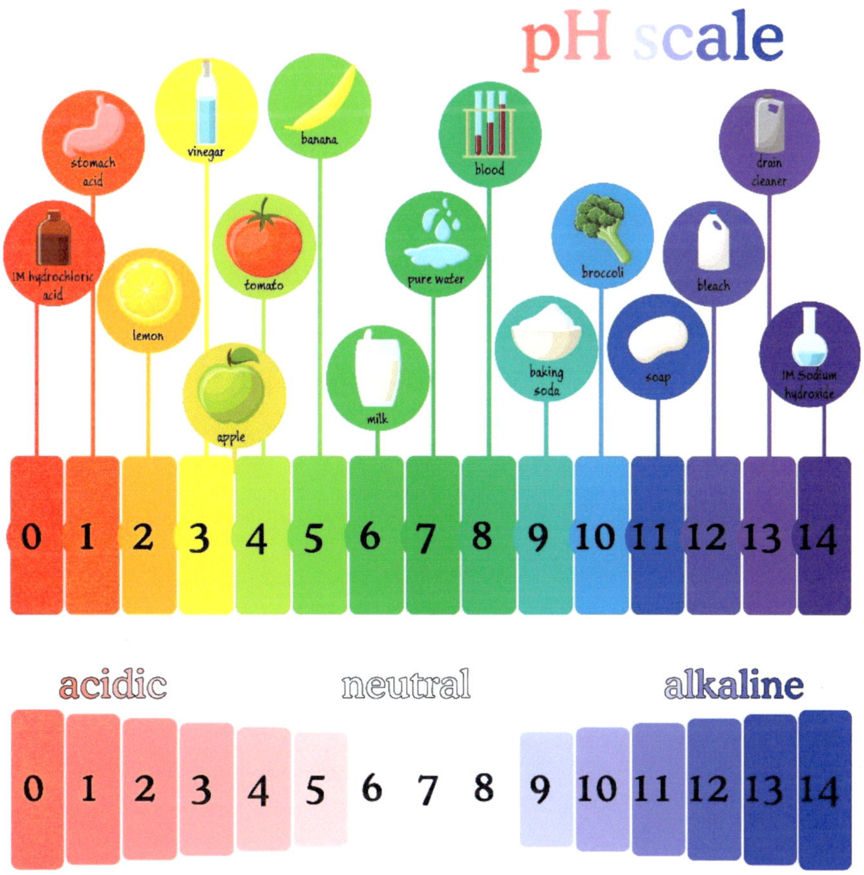

4.1.2 Neutralisation

At the end of this topic you should know:

- Acids are neutralised by alkalis (e.g. sodium hydroxide) and bases (e.g. insoluble metal oxides) to produce salts and water.
- Acids are neutralised by carbonates to produce salts, water and carbon dioxide.
- Be able to complete word equations for these reactions, given the names of the reactants.
- Carbon dioxide turns limewater milky.
- Salt solutions can be crystallised to produce solid salts.

Definition of terms

Acid – a chemical that contains hydrogen and has a pH less than seven.

Base – a chemical that can combine with an acid to produce a salt.

Alkali – a base that dissolves in water. Alkalis have a pH of more than seven.

pH – a scale that shows how acidic or alkaline a substance is.

Neutralisation.

Acids and bases react together to make salt and water. The solution formed is neutral. This is called neutralisation.

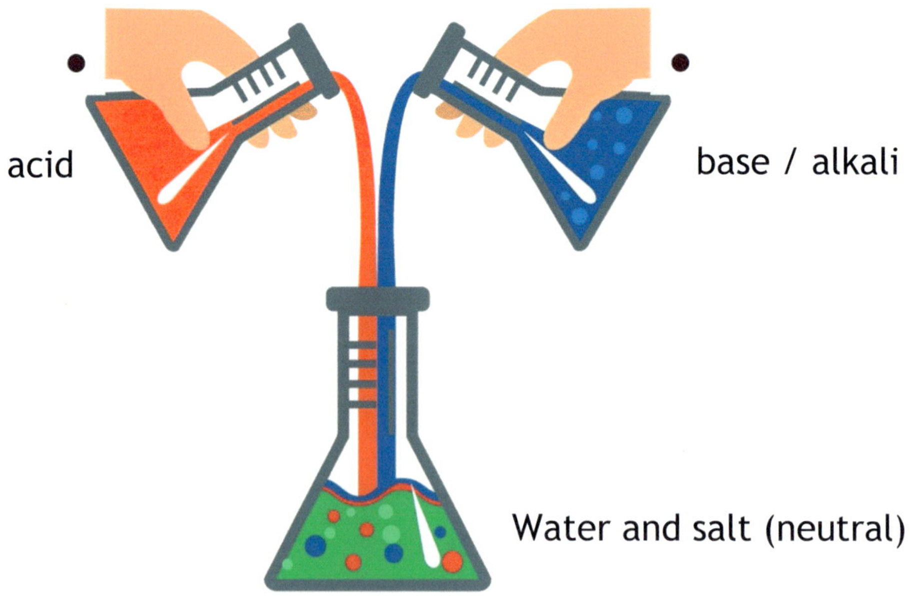

General Equation

acid + base → salt + water

This can also be written as

acid + alkali → salt + water

Bases and alkalis include:

- Metal hydroxides
- Metal oxides

For example. If magnesium oxide is reacted with sulfuric acid, then magnesium sulfate and water will be produced. (Remember the salt is magnesium oxide, because the metal we start with is magnesium and sulfuric acid always makes a sulfate). This can be written as:

Magnesium oxide + sulfuric acid → magnesium sulfate + water

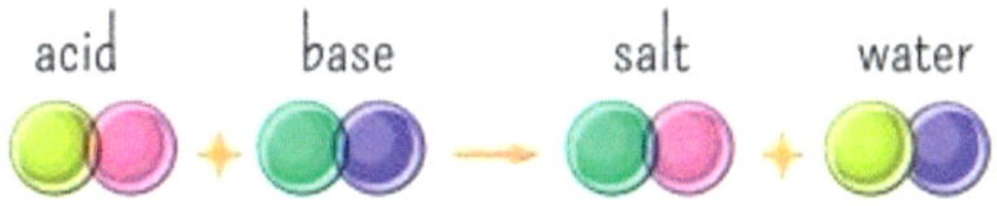

Metal Carbonates

Metal carbonates are also bases. An example is calcium carbonate which is limestone or marble.

Metal carbonates contain carbon, so when they react carbon dioxide is also formed.

General Equation

acid + metal carbonate → salt + water + carbon dioxide

This reaction is still a neutralisation reaction. For example, if sodium carbonate reacts with hydrochloric acid then sodium chloride is formed with water and carbon dioxide. (Remember the salt is sodium chloride, because the metal we start with is sodium and hydrochloric acid always makes a chloride). This can be written as:

sodium carbonate + hydrochloric acid → sodium chloride + water + carbon dioxide

Uses of Neutralisation

Your stomach contains hydrochloric acid, and too much causes indigestion. Antacid tablets contain bases which neutralise the extra acid.

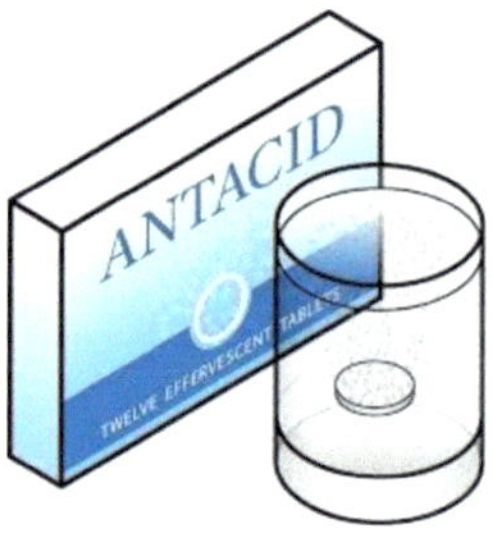

Bee stings are acidic, while wasp stings are alkaline. So, bee stings can be neutralised by using a weak alkali while to neutralise a wasp sting a weak acid could be used.

Insect	pH of sting	Treat with (to neutralise sting)	Example of what to treat with
Bee 	acidic	weak alkali	bicarb soda
Wasp 	alkali	weak acid	vinegar

Limewater Test

Limewater is a clear liquid. If carbon dioxide is bubbled through limewater, the carbon dioxide turns the limewater milky.

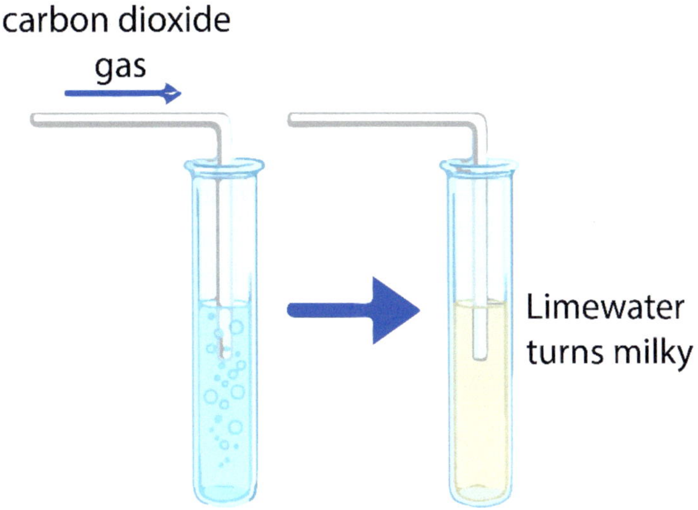

Crystallisation

Once the salt is formed it can be removed from the solution by evaporation. The water evaporates and leaves the salt behind.

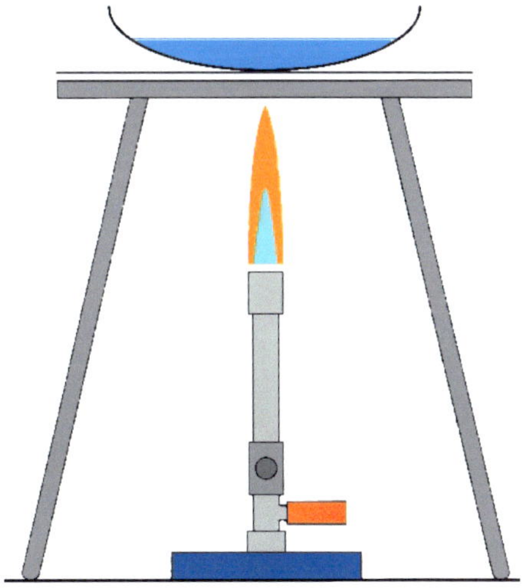

4.2 Energy and Rate of Reaction

4.2.1 Energy of Reactions

At the end of this topic you should know:

- Some reactions transfer energy to the surroundings so the temperature increases. Such reactions include combustion, oxidation and neutralisation.
- Other reactions take in energy from the surroundings, so the temperature decreases. These reactions include dissolving ammonium chloride in water and reacting citric acid with sodium hydrogencarbonate.
- The rate of a chemical reaction may be increased by increasing the temperature, increasing the concentration of reactants, increasing the surface area of solid reactants or by adding a suitable catalyst.

All chemical reactions either:

- Give out energy
- Take in energy

Reactions that give out energy

Some reactions transfer energy to the environment. This energy is transferred as heat energy. Therefore, the temperature of the reaction increases. This means that the temperature of the reaction gets hotter. Heat energy is called thermal energy.

Reactions that transfer heat energy to the surroundings includes:

- Combustion reactions. Burning coal is an example of a combustion reaction.
- Oxidation reactions. Iron rusting is an example of an oxidation reaction.
- Neutralisation.

A combustion reaction

Reactions that take in energy

Some reactions take in energy from the environment. Therefore, the temperature of the reaction decreases. This means that the temperature of the reaction gets colder.

Reactions that take in energy from the surroundings includes:

- Dissolving ammonium chloride in water
- Reacting citric acid with sodium hydrogencarbonate.

This is how disposable cold packs work.

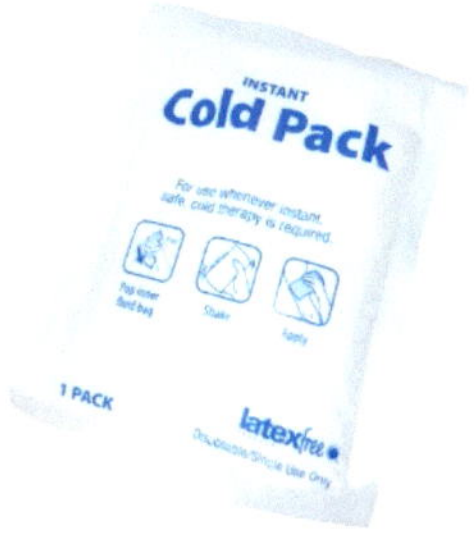

Reaction Rate

Reactions happen at different rates. Some reactions such as lighting a match happen very quickly while other reactions such as a nail rusting can take months or even years.

Which one is reacting the fastest? How can you tell?

We can speed up the rate of a reaction by:

- Increasing the temperature
- Increasing the concentration of the reactants
- Increasing the surface area of solid reactants
- Adding a catalyst

A catalyst is a chemical that increases the speed of a reaction but is not changed by the reaction.

A catalyst inside living organisms is called an enzyme.

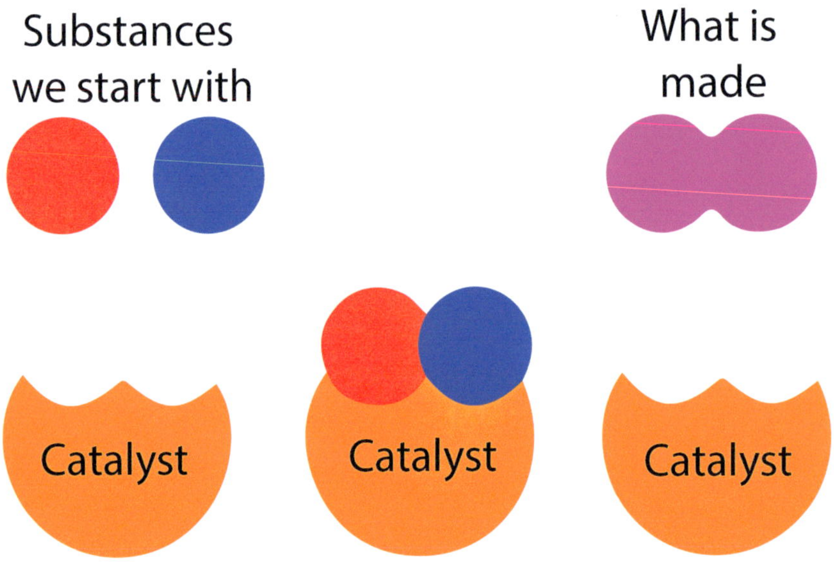

A catalyst is one way to speed up a reaction.

Measuring reaction rate

Reaction rate can be measured by either measuring:

- The substances you start with and see how much is used
- The substances formed and see how much is made.

In reactions where a gas is produced you can either:

- Count the bubbles produced
- Collect the gas and measure the volume.

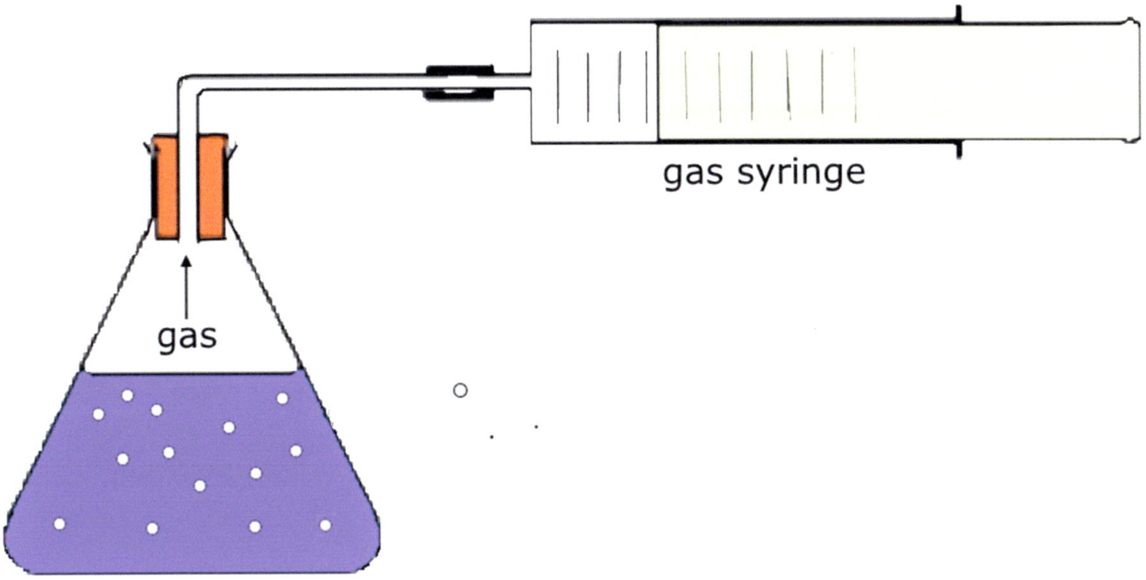

In reactions where a precipitate is formed, a cross can be placed under the reaction and the time taken for the cross to disappear measured.

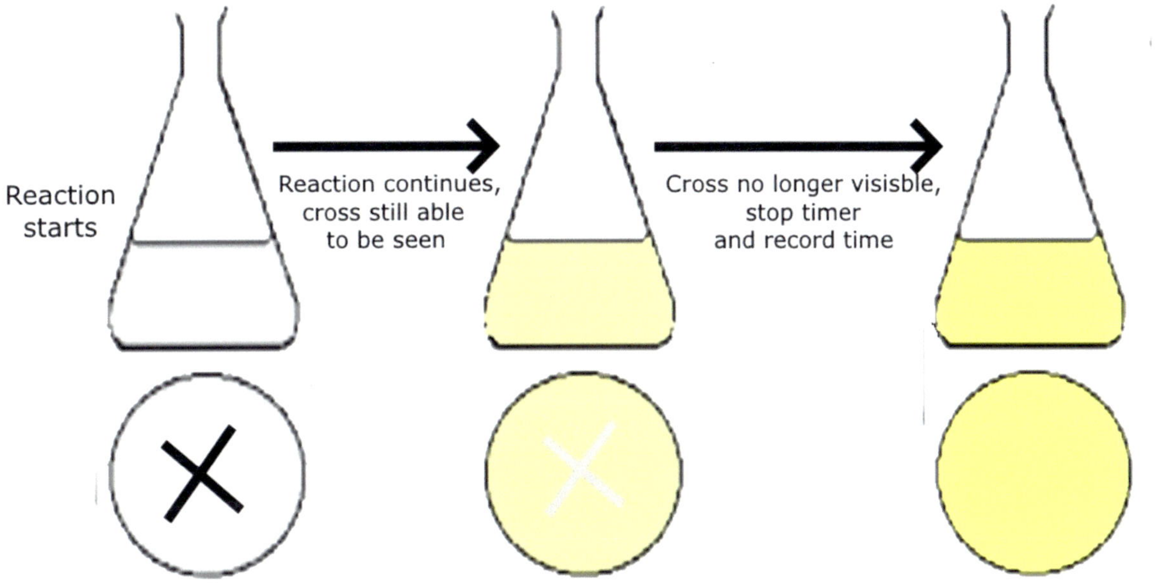

4.3 Earth's Atmosphere

4.3.1 The Earth's Atmosphere

At the end of this topic you should know:

- During the first billion years of the Earth's existence, there was intense volcanic activity that released gases that formed the early atmosphere and water vapour that condensed to form the oceans.
- The early atmosphere was mainly carbon dioxide with little or no oxygen.
- From about three billion years ago, algae and plants developed and produced the oxygen that is now in the atmosphere, by a process called photosynthesis.
- Photosynthesis can be represented by the word equation:
  carbon dioxide + water → glucose + oxygen.
- Carbon dioxide was removed from the early atmosphere by dissolving in the oceans and by photosynthesis. Most of the carbon from the carbon dioxide gradually became locked up in rocks as carbonates and fossil fuels.
- The Earth's atmosphere is now about four- fifths (80%) nitrogen and about one-fifth (20%) oxygen, with small amounts of other gases, including carbon dioxide, water vapour and argon, which is a noble gas.

The Earth's beginnings

While we cannot know exactly what the early Earth was like, or how old it is. This section looks at the current theory.

The atmosphere is the layer of gasses that surround a planet, such as the Earth.

On the early Earth there were a lot of volcanoes. The volcanic activity created the early atmosphere.

The early atmosphere was mainly carbon dioxide. There was some water vapour but no oxygen.

When the Earth cooled the water vapour condensed and made oceans.

The carbon dioxide in the early atmosphere was reduced by:

- Photosynthesis by algae and early plants.
- Carbon dioxide dissolving in the ocean.
- Carbon also became a part of rocks, as carbonates. The carbon in rocks would stay there for a very long time, so they are called carbon sinks.
- Another carbon sink is fossil fuels where carbon is locked up for thousands even millions of years.

Photosynthesis is the process that plants use to produce glucose and other molecules. In photosynthesis plants and algae combine carbon dioxide with water to produce glucose and oxygen.

The general equation for photosynthesis is:

carbon dioxide + water → glucose + oxygen.

In this way, photosynthesis both reduced the amount of carbon in the atmosphere and increased the amount of oxygen.

Today the Earth's atmosphere consists of:

- 78% Nitrogen
- 21% Oxygen
- 1% other gases including carbon dioxide.

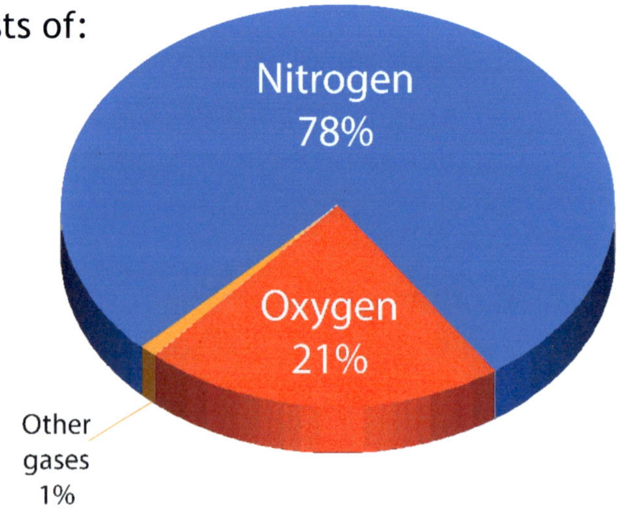

4.4 Fuels and Human Impacts on the Atmosphere

4.4.1 Crude Oil

At the end of this topic you should know:

- Crude oil is a mixture of a very large number of compounds. Crude oil is found in deposits underground, e.g. the oil fields under the North Sea.
- Crude oil may be separated into fractions by fractional distillation. This process, which takes place in a refinery, can be used to produce a range of useful fuels and oils, including petrol and diesel.

Crude Oil

Crude oil is a type of fossil fuel, found in underground deposits. It is made from dead plants and animals that dropped to the bottom of the sea millions of years ago. Crude oil is a mixture of lots of different compounds, called hydrocarbons. Most of the crude oil that we use in the UK comes from deposits in the North Sea.

Fractional Distillation

The different compounds found in crude oil have different properties and are used for different things. Crude oil as a mixture of these compounds is not useful, but crude oil is separated in a process called fractional distillation into fractions. These fractions are very useful.

Fractional distillation works because the different compounds found in crude oil have different boiling points. The oil is heated to a very high temperature and then cooled. Then compounds with similar boiling points are collected. The compounds that are collected together are called fractions. Fractional distillation happens in an oil refinery.

Some of the fractions of crude oil include:

- Gas
- Petrol
- Naphtha - we use this to make plastics
- Diesel
- Aircraft fuel
- Bitumen - we use this to surface our roads.

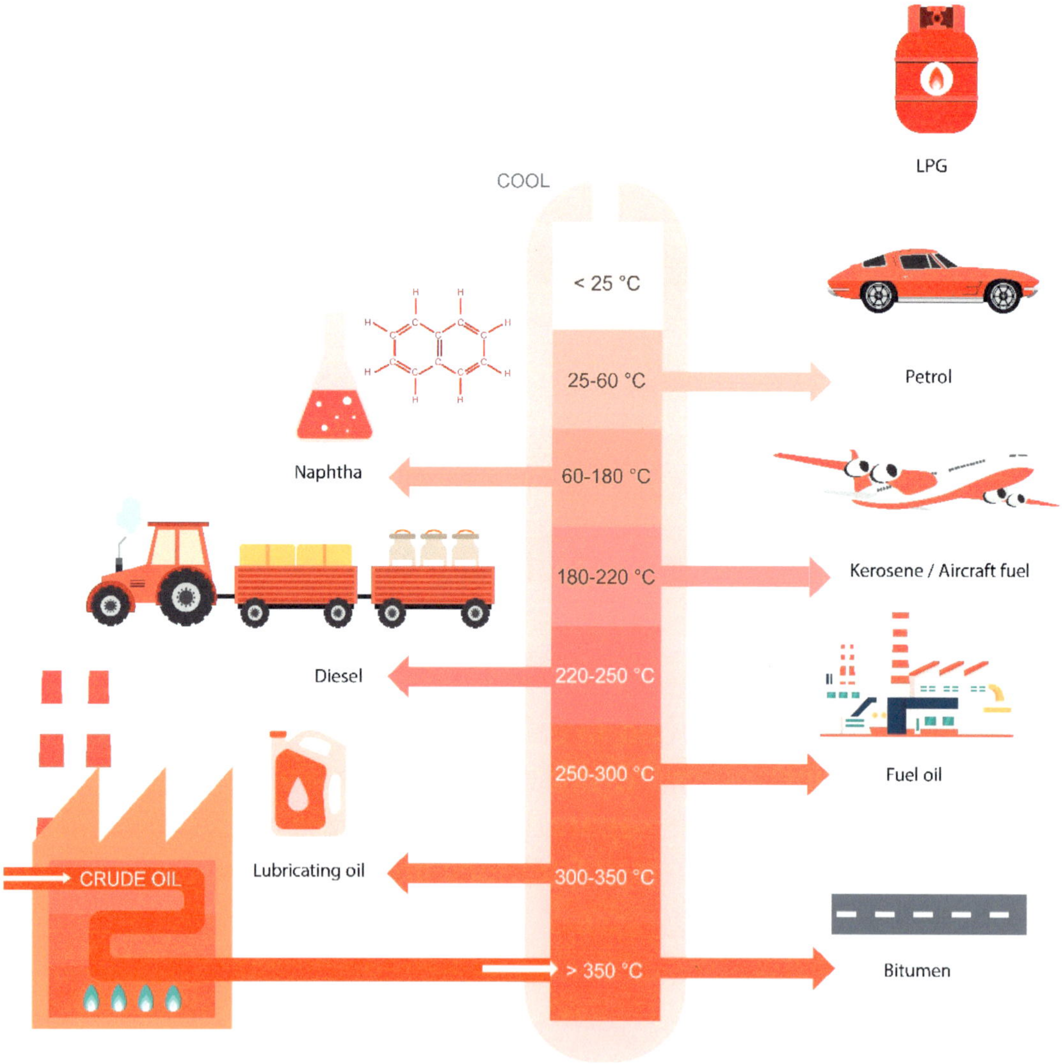

4.4.2 Burning Fossil Fuels

At the end of this topic you should know:

- When fuels burn completely the gases released into the atmosphere include carbon dioxide, water (vapour), and oxides of nitrogen. Sulfur dioxide is also released if the fuel contains sulfur.
- When fuels burn in a limited supply of air carbon monoxide is produced. Solid particles (soot) may also be produced.
- Be able to describe the impact on the environment of burning fossil fuels.
- Oxides of nitrogen and sulfur dioxide cause acid rain and problems for human health.
- Carbon monoxide is a colourless, odourless, poisonous gas that can cause death.
- Solid particles can cause global dimming and problems for human health.

Fossil fuels can be burnt completely. That means that there is enough oxygen for the carbon being burnt to combine with the oxygen and make carbon dioxide.

When fossil fuels are burnt completely gases are produced including:

- Carbon dioxide
- Water vapour
- Nitrogen oxides
- Sulfur dioxide (not always)

Sulfur dioxide and nitrogen dioxide dissolve in water vapour in the air and cause acid rain. Acid rain causes damage to rocks, buildings, statues, lakes, plants and animals.

Forest destroyed by acid rain.

Sulfur dioxide also:

- Dissolves in lakes and rivers making them acidic
- Kills plants
- Can be breathed in and damage human health.
- Damages buildings

Incomplete Combustion

When a fuel is burnt where there is not enough oxygen to convert all of the carbon to carbon dioxide, carbon monoxide and soot can be made instead.

The equation for incomplete combustion is:

fuel + oxygen → carbon monoxide + soot + water.

Carbon monoxide is a molecule of a carbon atom with one atom oxygen, compared to two atoms of oxygen in carbon dioxide.

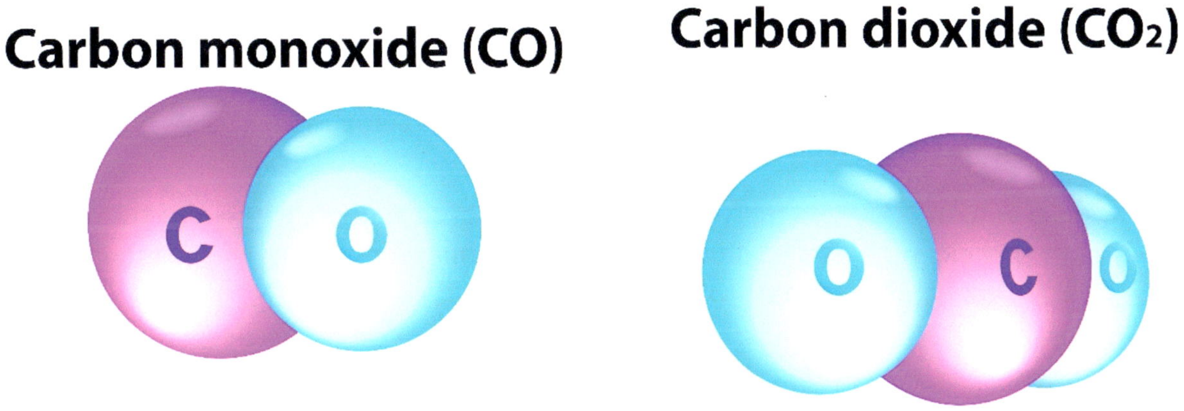

Carbon monoxide is:

- Colourless
- Odourless
- Poisonous
- Kills people and animals

This makes carbon monoxide very dangerous because people can't see it or smell it, but it can kill them. This is why gas boilers need to be serviced.

Soot is powdered carbon. When people breathe in soot it can cause heart and breathing problems. It also causes everything to become dirty.

Soot and fine solid particles that can be produced in the burning of fossil fuels can block sunlight reducing the amount of light reaching the Earth's surface. This is called Global Dimming.

4.4. Greenhouse Gases

At the end of this topic you should know:

- Some human activities increase the amounts of greenhouse gases in the atmosphere, such as carbon dioxide from burning fossil fuels and methane from landfill and cattle farming.
- Increased levels of greenhouse gases in the atmosphere cause the temperature to increase. Many scientists believe that this will result in global climate change.

Greenhouse gases include carbon dioxide and methane. Human activity is increasing the amount of these gases in the air.

- Carbon dioxide is released into the atmosphere when fossil fuels are burnt.
- Methane is made when the rubbish in landfills breaks down and also from cows on farms.

The greenhouse gases in the atmosphere prevent heat from escaping.

The heat energy from the Earth is reflected by the greenhouse gases back to the Earth.

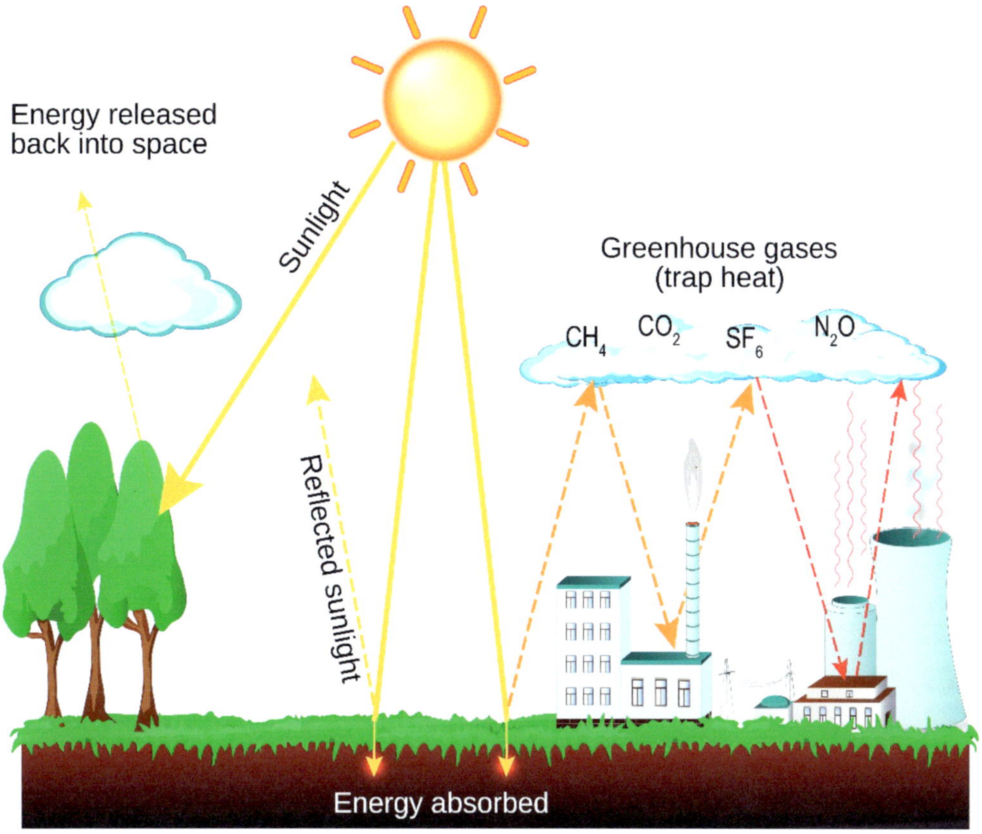

Global warming is causing:

- Temperatures to rise
- Sea levels to rise
- Extreme weather, such as tsunamis and droughts.

4.5 Water for Drinking

4.5.1 Drinking Water

At the end of this topic you should know:

- Water that is safe to drink has small amounts of dissolved substances and low levels of microbes.
- Most drinking water is produced by choosing a suitable source of fresh water, filtering to remove solids and sterilising to kill microbes.
- If supplies of fresh water are limited, salty water can be distilled to produce fresh water. This requires a large energy input.

Safe water

Having safe water is very important as unsafe water causes disease and can even kill people, particularly children.

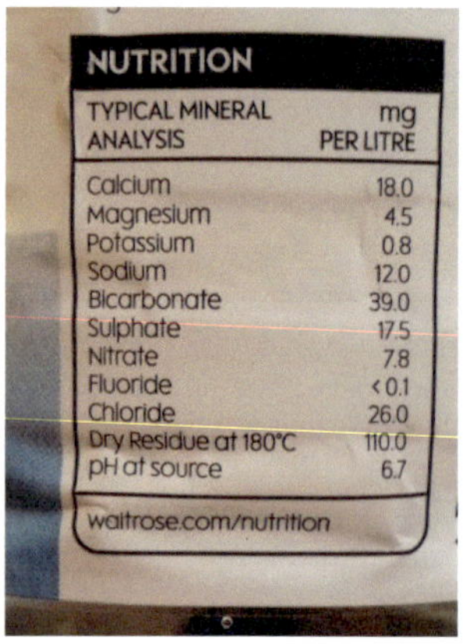

Water that is safe to drink has small amounts of dissolved substances, such as salts, and low levels of safe microbes.

Using fresh water

Fresh water is water that does not have much salt in it. It is the opposite of seawater.

To make fresh water safe:

- Filter to remove solids
- Sterilise to kill microbes.

After this the water may still contain:

- Salts
- Chlorine
- Dead microbes

Using salty water

If there is a lack of fresh water, then salty water can be used.

To make fresh water from salty water, distillation is used. The problem with this is it uses a lot of energy.

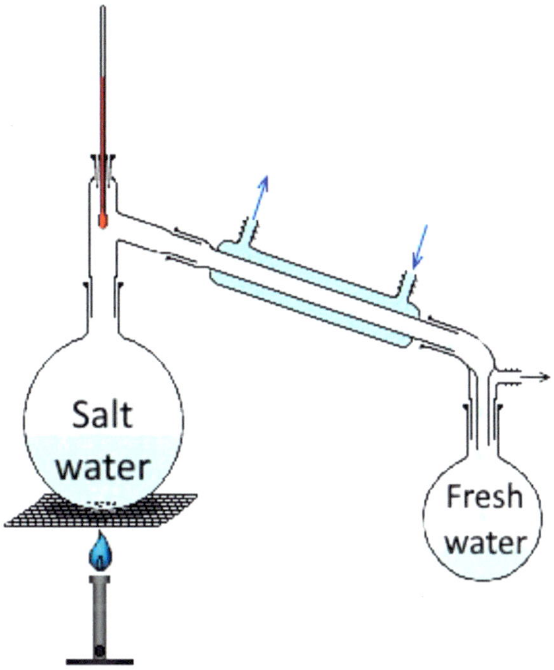

Physics Component 5
Energy, Forces and the Structure of Matter

5.1 Energy, Energy Transfers and Energy Resources

5.1.1 Energy

At the end of this topic you should know:

- Be able to describe different energy stores. In particular, kinetic, gravitational potential, thermal, light, sound and elastic (strain).
- Describe, for common situations, the changes involved in the way energy is stored when a system changes. For example:
    - an object projected upwards
    - a moving object hitting an obstacle
    - a vehicle slowing down
    - bringing water to a boil in an electric kettle.
- Be able to identify and describe useful and wasted energy.
- Energy can be transferred usefully, stored or dissipated, but cannot be created or destroyed.
- Energy efficiency refers to the amount of useful energy compared to the total energy.
- Whenever there are energy transfers in a system only part of the energy is usefully transferred. The rest of the energy is dissipated so that it is stored in less useful ways. This energy is often described as being 'wasted'.
- Unwanted energy transfers can be reduced in a number of ways, e.g. through lubrication and the use of thermal insulation.
- How the rate of cooling of a building is affected by the thickness and thermal conductivity of its walls.
- The higher the thermal conductivity of a material the higher the rate of energy transfer by conduction across the material.

Energy is the ability to do work. It is measure in Joules.

There are many different types of energy. If energy can be kept as that type for use later, it is called an energy store. The main types are:

- Kinetic
- Gravitational potential
- Chemical
- Light
- Sound
- Thermal
- Elastic (strain)

Kinetic

Anything that moves has kinetic energy.

The amount of kinetic energy depends on the speed and the mass.

Examples include a bouncing ball, a moving car or a planet.

Gravitational Potential

Gravitational potential energy is the energy due to position. The higher an object is above the ground the more gravitational potential energy it has.

Examples include: a ball about to be dropped or a book on a desk.

Chemical

Chemical energy is the energy stored in the bonds of chemicals.

Examples include: fuel, food and batteries.

Light and Sound

Sound is created by objects vibrating. Light can have different colours. While these are types of energy, they are not energy stores. Examples include a television or a computer.

Thermal

Thermal energy is heat energy. The hotter something is the more thermal energy it has.

An example would be a hot cup of coffee.

Elastic (strain)

The energy an object has when it is stretched or squashed.
An example is a stretched spring or rubber band.

Energy Transfer

Energy cannot be made or destroyed but can be transferred from one type to another.

For example, when a torch is turned on the chemical energy in the battery is converted to light energy.

There are four main ways energy can be transferred between stores. These are:

- by heating, due to a difference in temperature.
- mechanically, a force moving an object, for example by lifting a ball from the ground.
- electrically, by the movement of charges.
- by light or sound

When a ball is thrown upward, the gravitational potential energy will increase, but the kinetic energy will decrease. At the top of the throw, the balls speed will be zero, but it will have the most gravitational potential energy. As it falls back down the kinetic energy will increase as its speed increases, but the gravitational potential energy will again decrease.

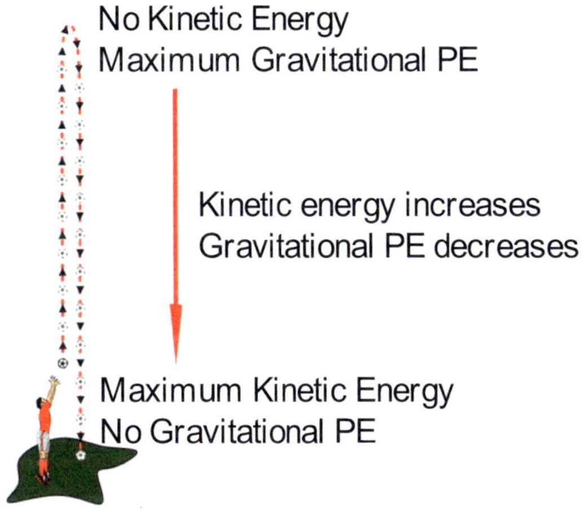

When a kettle boils the thermal energy increases, as the water gets hot. The water moves around, this is kinetic energy.

Waste Energy

When energy is transferred from one form to another, not all of the energy is useful. Some energy is also changed into forms that we do not want. This is called waste energy.

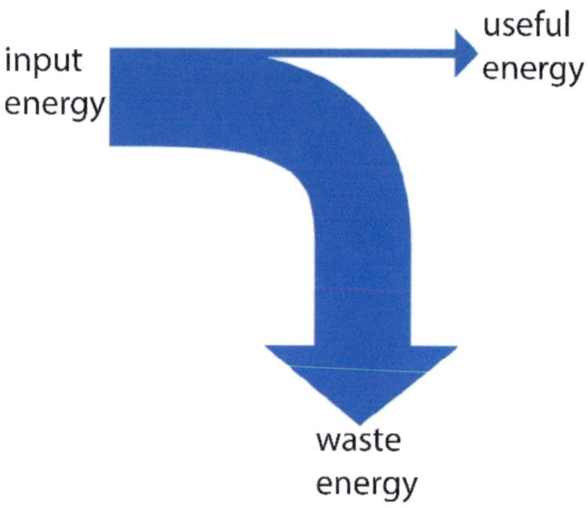

For example, when we turn on a torch for light, some of the energy also becomes heat energy.

When a vehicle slows down, the speed decreases and therefore the kinetic energy decreases. This energy is lost as sound and heat.

Light, heat and sound energy spread out. When energy spreads out, we say it dissipates. The dissipated energy is not useful.

Energy Efficiency

When energy is transferred the greater the proportion of energy that is useful, the more efficient the transfer.

Energy efficient lightbulb. Most energy becomes light.

Energy inefficient lightbulb. Most energy is dissipated as heat. A little energy becomes light.

Heat always moves from where there is more heat to where there is less heat.

The amount of energy that spreads out through the surroundings can be reduced. This means that we need less energy to produce the same amount of useful energy. This means that we make it more efficient.

Thermal Conduction

One-way heat energy is transferred is through conduction. Conduction is the transfer through solids. It is caused by particles in the solid vibrating and bumping into neighbouring particles.

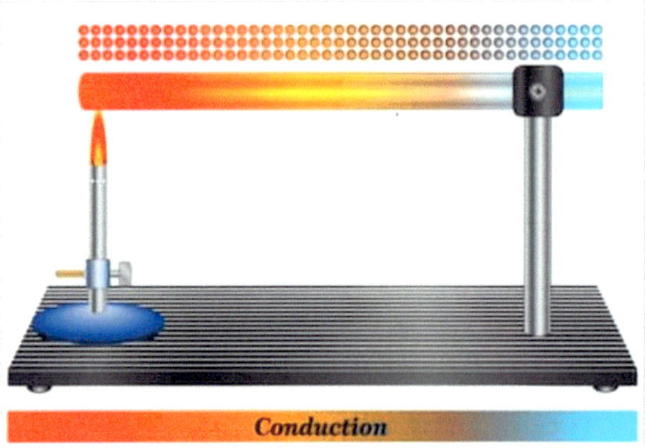

Thermal conductivity refers to how quickly heat energy is transferred by conduction through a material.

Metals conduct very well.

Insulation

Substances that don't conduct well, that is they have a low thermal conductivity, are called insulators. Insulators can be used to prevent heat being lost to the surroundings.

How well an insulator works depends on:

- The thermal conductivity of the material
- The thickness of the material.

Insulating Houses

Here is how the heat is lost in a typical uninsulated home.

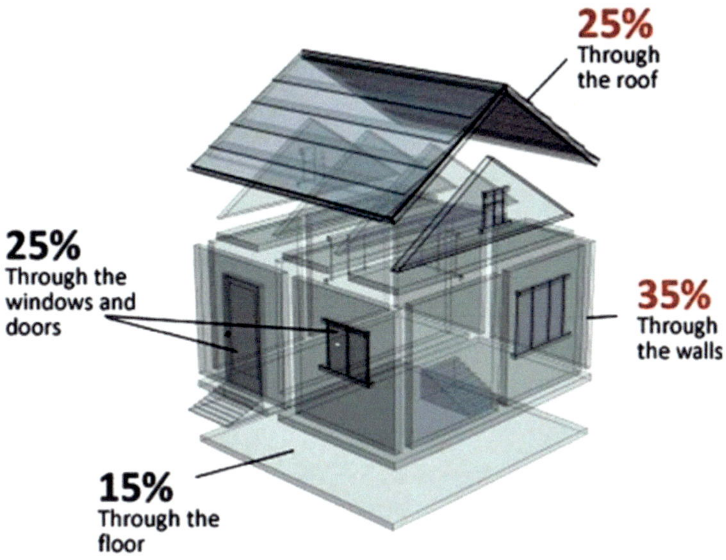

The amount of heat lost in a house can be reduced in a number of ways, including:

- Loft insulation
- Cavity-wall insulation
- Double glazing
- Carpets and curtains.

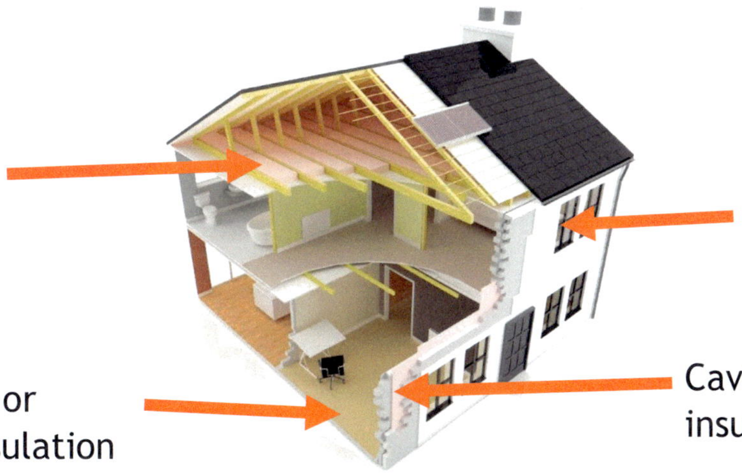

5.1.2 Energy Resources

At the end of this topic you should know:

- Describe the main energy resources available for use on Earth. These include fossil fuels (coal, oil and gas), nuclear fuel, biofuel, wind, hydroelectricity, geothermal, the tides, the Sun, water waves.
- Distinguish between energy resources that are renewable and energy resources that are non-renewable.

Energy resources can be either renewable or non-renewable. A renewable energy resource can be replaced. However non-renewable cannot be replaced. They are made extremely slowly, so they will run out.

The main non-renewable energy resources are fossil fuels. Coal, oil and gas are fossil fuels. Fossil fuels are made from dead plants and animals that were buried a long time ago.

When fossil fuels are burnt, they produce carbon dioxide which increases global warming.

Nuclear power is also a non-renewable energy resource. Nuclear power is produced by splitting atoms. The nuclei give out ionising radiation in a process called radioactive decay. The heat produced from splitting the atom is used to heat water, producing steam. The steam turns turbines which generates electricity.

Renewable energy

Wind 	Wind turns the blades of the turbine. The turbine turns a generator producing electricity. Wind turbines involves no fuel costs and release no harmful gases. Newer turbines do not hurt birds. However, they only work when there is wind. They are noisy and may spoil the view for people nearby.

Water 	Water can also be used to drive turbines. This can happen in three ways: 1. Waves - The water in the sea rises and falls. Wave machines use this kinetic energy to drive generators. 2. Tides - Water from high tide is trapped behind a barrier. The water turns a turbine when it's released. 3. Hydroelectric also uses the kinetic energy of water to turn turbines. A dam is built across a river. The water above the dam has a lot of gravitational potential energy which is converted to kinetic energy when is rushes through tubes inside the dam or through the dam wall.
Solar 	Solar energy uses the energy from the sun to make electricity. It can do this either by heating water or by using solar cells to generate electricity. There are no fuel costs and no dangerous gases are released. Solar power can work in remote locations. However, solar cells are expensive and inefficient. Solar power does not work at night and does not work well in cloudy conditions.

Geothermal 	Hot water and steam from underground are used to drive turbines. The water may be from underground or cold water may be pumped underground where it is heated. Geothermal energy has no fuel costs and releases no dangerous gases. However, there are very few places suitable for producing geothermal energy.
Biofuels 	Biofuels are made from living things. Methane can be made from animal waste. Biodiesel is made from plant oils and bioethanol is made from fermenting sugar and wheat. The carbon absorbed in photosynthesis cancels out the carbon dioxide released when the fuel is burnt. However, the crops grown to produce biofuels may reduce the amount of farmland used to produce food.

These energy sources are used to make electricity. To make electricity a power station uses a turbine to turn a generator.

5.2.1 Forces and Work

At the end of this topic you should know:

- A force is a push or pull that acts on an object due to the interaction with another object.
- All forces between objects are either:
  - contact forces - the objects are physically touching
  - non-contact forces - the objects are physically separated.
- When a force causes an object to move through a distance, work is done on the object.
- Work done against the frictional forces acting on an object causes a rise in the temperature of the object.

A force is a push or a pull. A twist is also a force because it is a push or a pull at an angle.

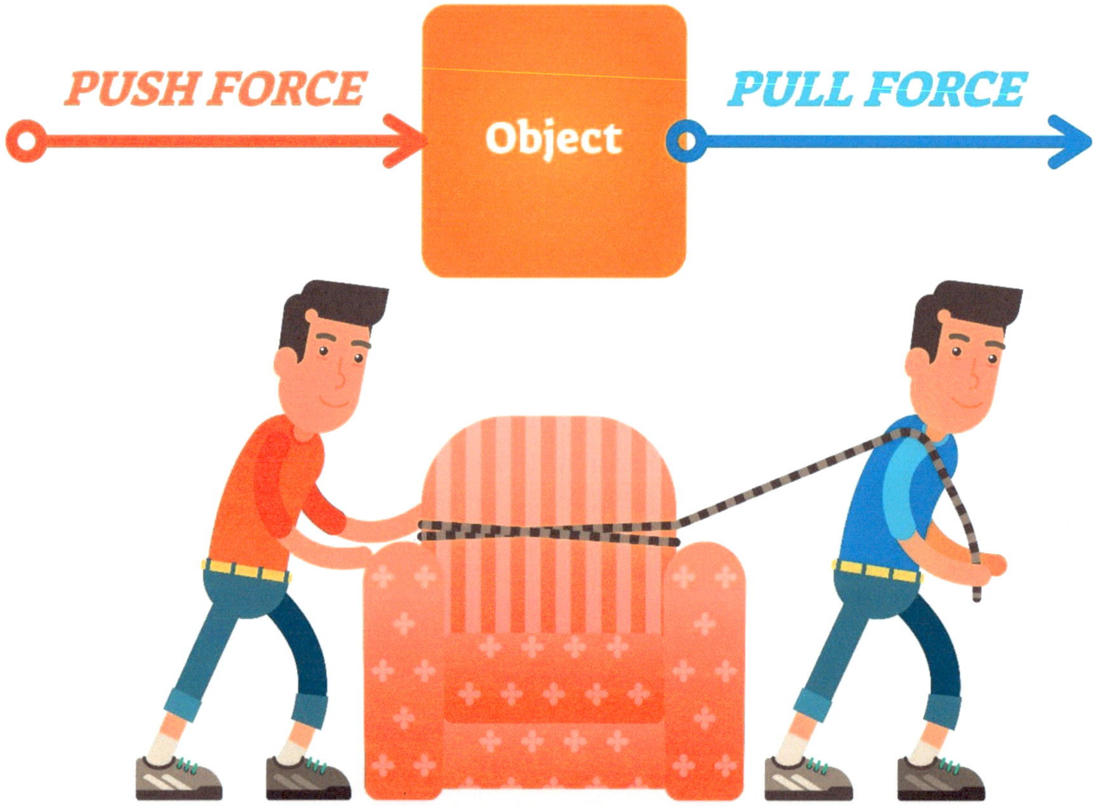

There are two types of forces – contact forces and non-contact forces.

Contact forces require the two objects to be touching, such as the friction of a car tyre on the road or pushing a door open.

Non-contact forces can happen without the objects touching, such as magnetism or gravity.

Contact forces	Non-contact forces
 Friction	 Gravitational force (Gravity)
 Air resistance	 Electrostatic force
 Tension	 Magnetism
 Normal contact force	

Friction

Friction is the force that opposes motion. Friction causes the objects to become hotter. Sometimes friction is useful, such as when we want to slow down or we want grip. Sometimes friction is unhelpful, such as when it causes unwanted heat, or wear and tear.

Friction is caused by two surfaces rubbing together. Even surfaces that appear really smooth are not completely smooth at a microscopic level.

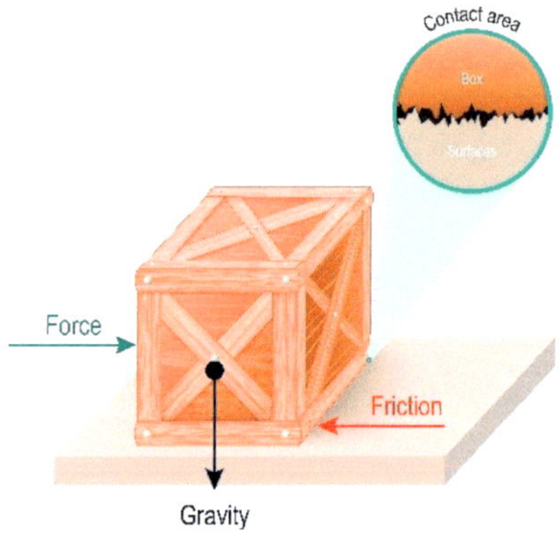

Static friction and sliding friction are greater than rolling friction because the surface area is bigger.

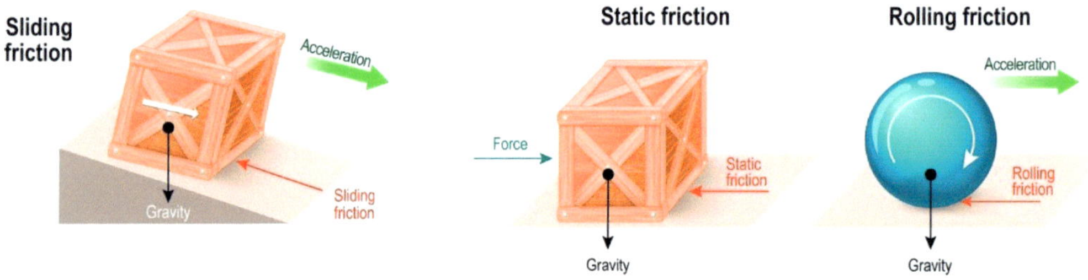

Work

When a force causes an object to move through a distance work is done on the object.

Work is measured in Joules, the same as energy. One joule of work is done when a one newton weight is raised by one metre.

The greater the force needed, or the greater the distance moved, the greater the work done.

5.3.1 Speed

At the end of this topic you should know:

- Speed is measured by the distance travelled in a certain time.
- Units of speed include metres per second, kilometres per hour and miles per hour.
- Be able to calculate average speed, using the equation speed = distance ÷ time

Speed is the distance travelled in a particular time.

It can be measured in km/h (kilometres per hour), mph (miles per hour) or m/s (metres per second). Per means each or divided by, so if people are travelling at 60km/h means that they are travelling 60 km each hour.

Speed can be calculated using the formula:

$$speed = \frac{distance}{time}$$

So, if a car travels 140 miles in 4 hours we can calculate the speed as follows:

1. Write down the formula.

$$speed = \frac{distance}{time}$$

2. Substitute in the variables you know.

$$speed = \frac{140}{4}$$

3. Do the calculation. Remember in Science you can always use a calculator.

$$\frac{140}{4} = 35$$

4. Write down your answer. Remember the units.

35 mph

5.3.2 Stopping Distances

At the end of this topic you should know:

- The stopping distance of a vehicle is the sum of the distance the vehicle travels during the driver's reaction time (thinking distance) and the distance it travels under the braking force (braking distance).
- For a given braking force the greater the speed of the vehicle, the greater the stopping distance.

A moving vehicle takes time to stop. The time it takes to stop is called the stopping distance.

The distance it travels while the driver is deciding to stop is called the thinking distance and the distance it travels while it is decelerating (slowing down) is called the braking distance. The stopping distance is the total distance taken (thinking distance + braking distance)

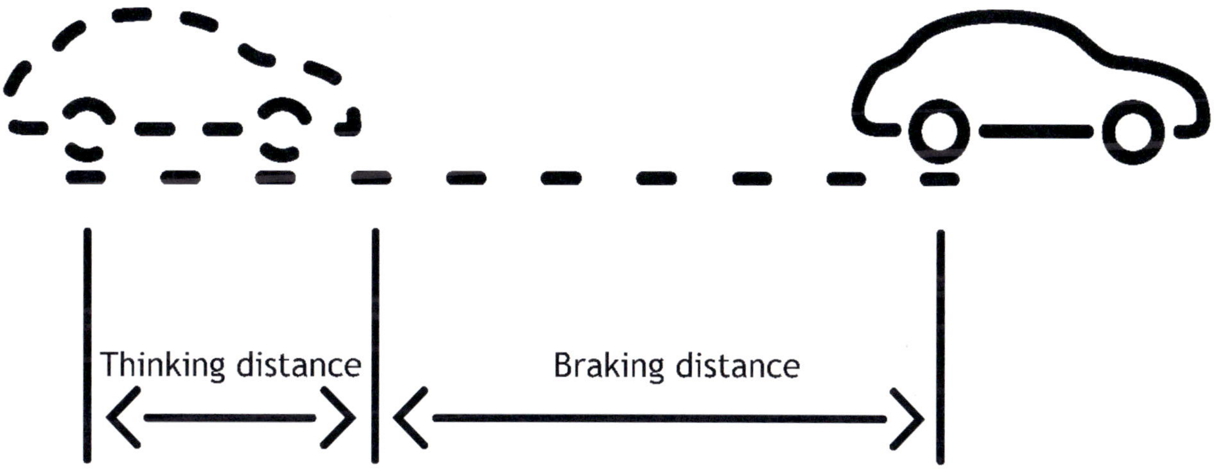

Therefore, stopping distance can be calculated using the formula:

$$stopping\ distance = thinking\ distance + braking\ distance$$

For any particular vehicle, the stopping distance increases with increasing speed. At increased speed both the thinking distance and the braking distance is increased.

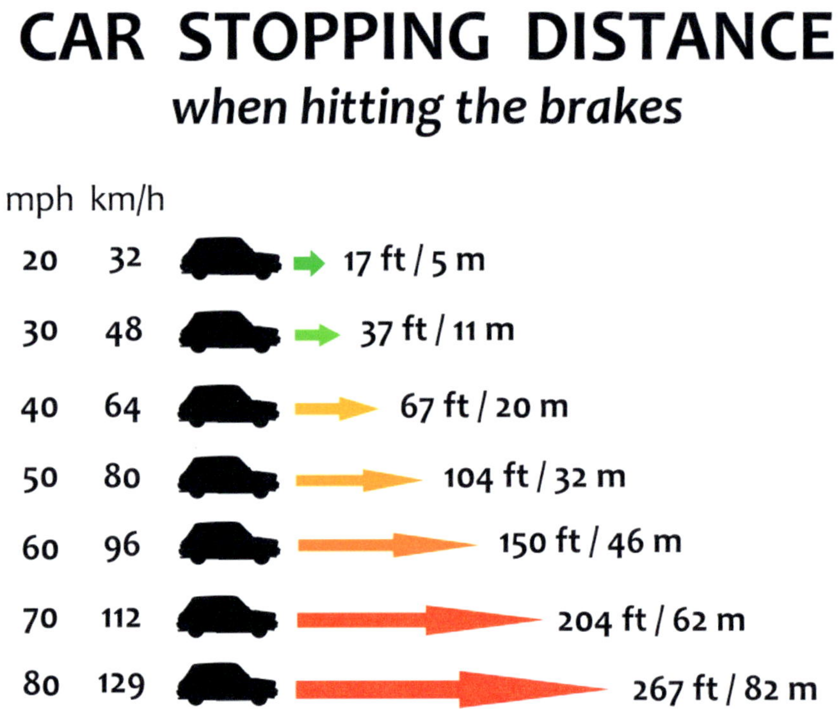

This means that people need to leave a greater distance between vehicles when travelling faster, to avoid accidents.

5.3.3 Factors affecting stopping distance

At the end of this topic you should know:

- Reaction times vary from person to person. Typical values range from 0.5 s to 0.9 s.
- Different methods for calculating reaction time.
- How a driver's reaction time can be affected by tiredness, drugs and alcohol. Distractions may also affect a driver's ability to react.
- The braking distance of a vehicle can be affected by adverse road and weather conditions, and poor condition of the vehicle.
- Be able to analyse a given situation to identify how braking could be affected.

Reaction Time

A person's reaction time is how quickly they respond to something.

Reaction times vary from person to person, but are typically 0.5 seconds (s) to 0.9 s.

A driver's reaction time can be affected by:

- tiredness
- drugs
- alcohol
- distractions

Longer reaction times increase the thinking distance when stopping from a given speed.

There are different ways to measure reaction times. One simple method involves dropping a ruler between someone's open thumb and forefinger. The further the ruler falls before being stopped, the higher the reaction time of the person grabbing the ruler. The ruler can be marked in s rather than in mm.

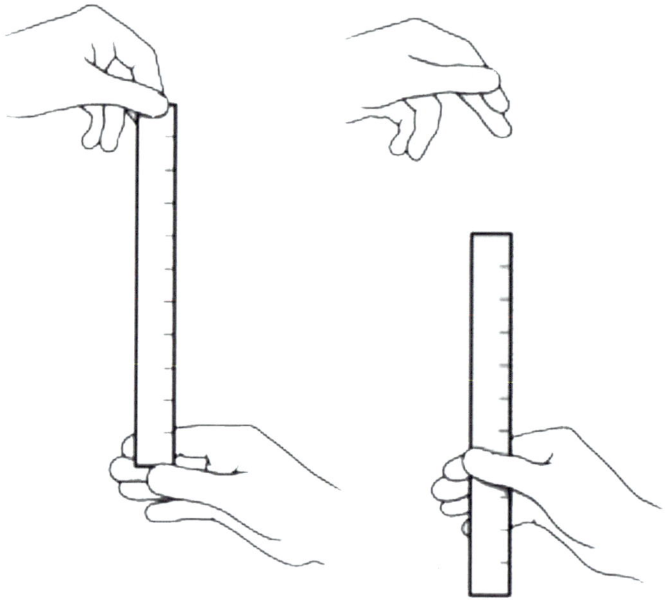

Another common way to measure reaction time is by using a website that asks the user to press a button when something happens, such as change in colour.

The braking distance of a vehicle can be affected by adverse weather conditions and poor condition of the vehicle, particularly the brakes.

5.4.1 Atoms and Nuclear Radiation

At the end of this topic you should know:

- Some atomic nuclei are unstable.
- The nucleus gives out ionising radiation.
- This is a random process called radioactive decay.
- The nuclear radiation emitted may be:
    - alpha particles
    - beta particles
    - gamma rays.
- Properties of alpha particles, beta particles and gamma rays limited to their penetration through materials and their range in air.
- Know some of the uses and dangers of the three types of radiation.

Atomic Nuclei

The nucleus is the centre of the atom. The plural of nucleus is nuclei. While most atoms are stable, some atoms have an unstable nucleus. The larger an atom the more likely the nucleus will be unstable.

Unstable atoms are radioactive because they atoms give out radiation to become more stable. When they do the atom loses one or more particles from its nucleus. This is called radioactive decay. When a particular atom emits radiation is random.

There are three types of radiation: alpha, beta and gamma.

- Alpha is a helium nucleus containing two protons and two neutrons.
- Beta is an electron.
- Gamma is a high energy wave. It does not have a charge.

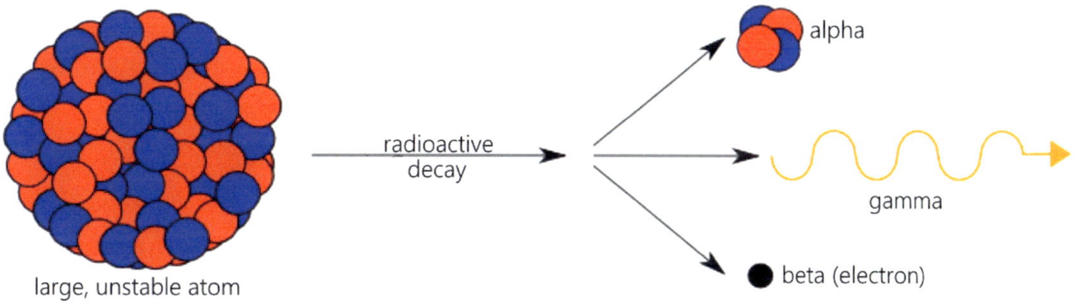

These three types of radiation have different properties. One difference is how far they can travel through different materials. In air alpha particles can travel only 1-3 cm, while beta particles can travel much further and gamma particles can travel the furthest. Alpha is blocked by a sheet of paper or even air, beta radiation can be blocked by just a few millimetres of aluminium. However, gamma radiation can only be blocked by a thick layer of lead or concrete.

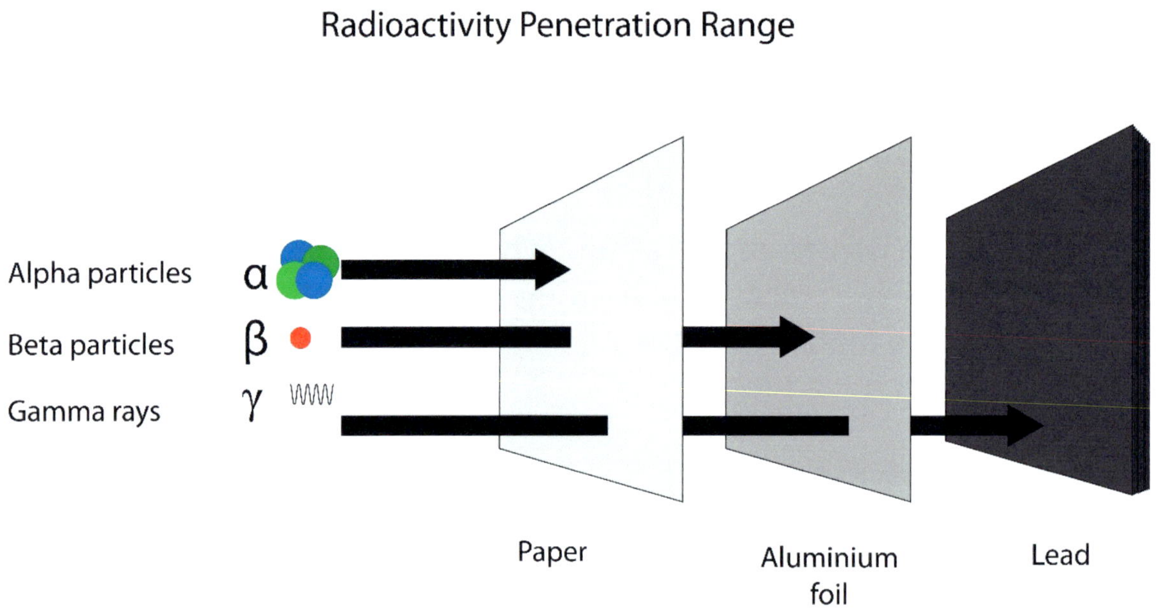

This doesn't mean that gamma radiation is always the most dangerous, as once inside the body, alpha particles do more damage.

Radiation can be useful. For example: alpha radiation is used in smoke alarms, beta radiation is used to keep some fruit fresh for longer and also to check the thickness of thin sheets of metal, and gamma radiation is often used in cancer treatment, by killing the cancer cells and also to sterilise surgical instruments.

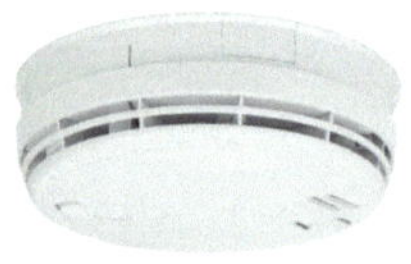

However, radiation can also be dangerous as it can cause cancer and other health problems.

Physics Component 6

Electricity, Magnetism and Waves

6.1 Electrical Current

6.1.1 Current

At the end of this topic you should know:

- Electric current is a flow of electrical charge. The size of the electric current is the rate of flow of electrical charge.
- The current through a component depends on both the resistance of the component and the voltage across the component. The current through a component depends on both the resistance of the component and the voltage across the component.
- The greater the resistance of the component the smaller the current for a given voltage across the component.
- The resistance of a component is a measure of how difficult it is for an electric current to pass through it.
- A complete circuit is necessary for a current to flow.
- Cells and batteries supply current that always passes in the same direction. This is called direct current (dc).
- An alternating current (ac) is one that changes direction.
- Mains electricity is an ac supply. In the UK it has a frequency of 50 Hz and is 230 V.

Current, Voltage and Resistance.

Electrical current is the flow (movement) of charges. In wires and electrical appliances, the charges that move are electrons.

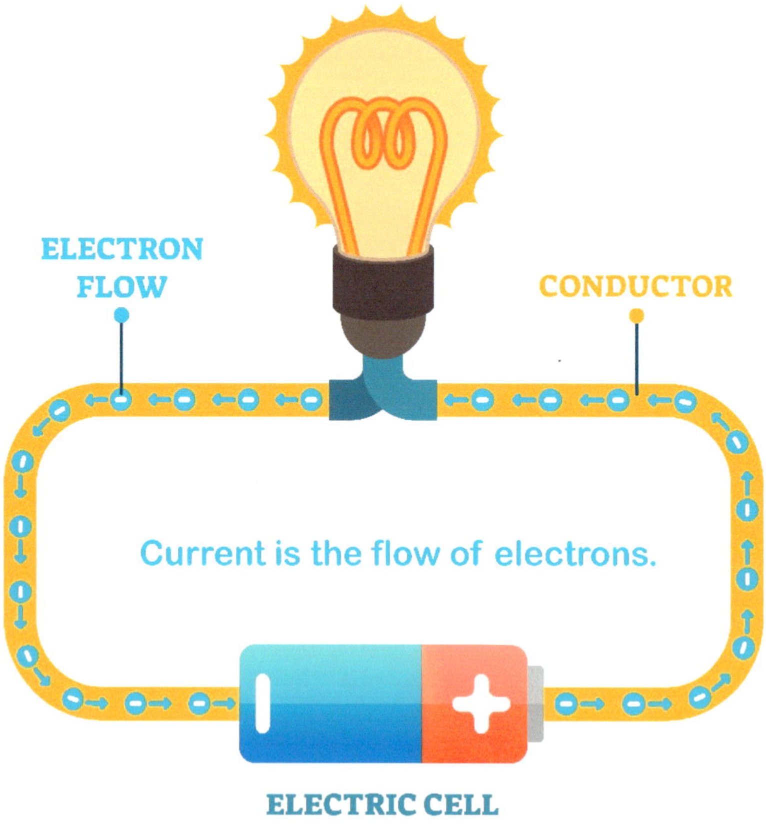

Current is measured in amps with an ammeter.

An amp is the number of charges that pass a particular point each second.

Voltage is the difference in energy between two points of a circuit.

Voltage is measured in volts with a voltmeter.

A volt is the number of joules of energy per charge.

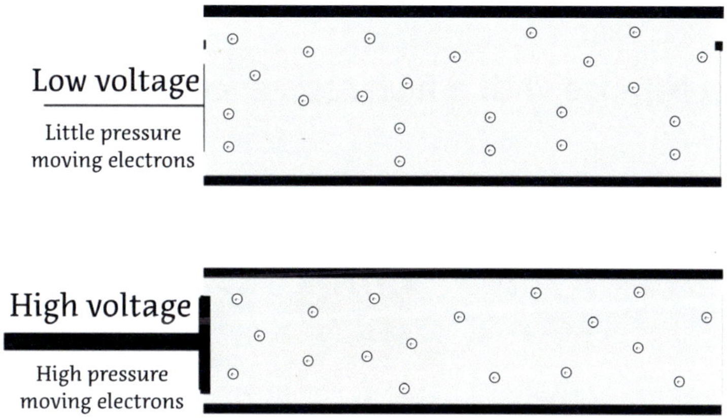

Resistance is how difficult it is for the current to flow through a conductor.

High resistance

Low resistance

The current is affected by both voltage and resistance.

- Increasing the voltage → increases the current.
- Increasing the resistance → decreases the current.

Increasing the voltage means that charges are getting a bigger push from the power supply and the charges have more energy. More energy means that they will move faster and therefor the flow will increase.

Increasing the resistance means that it is harder for the current to flow, so it will flow slower. Just like it is easier to run faster on flat ground than up a hill.

An insulator is a substance with a high resistance so little or no current can flow.

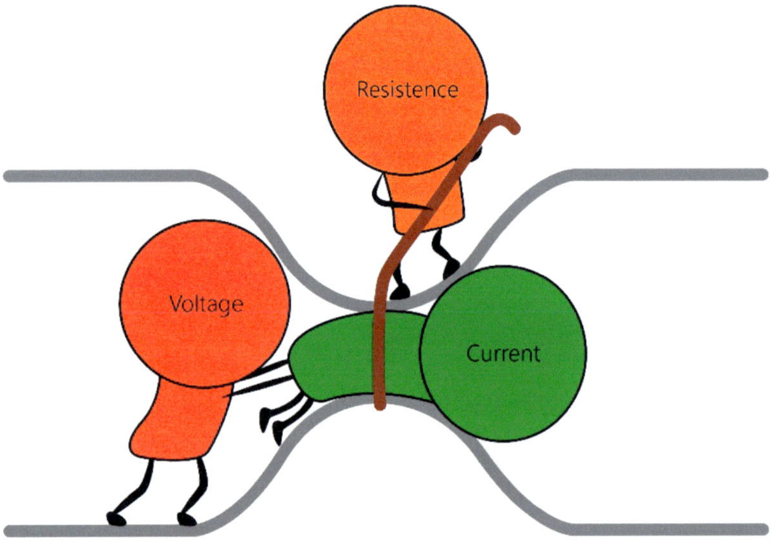

This diagram shows that the voltage is pushing, the resistance is making it harder for the current to move and the current is moving through the wire. Therefore, the voltage is causing the current to move faster while the resistance is slowing it down.

For current the charges need to flow. Therefore, there needs to be a complete circuit, as if the charges come to a "dead-end" they will be forced to stop, which will stop all of the charges.

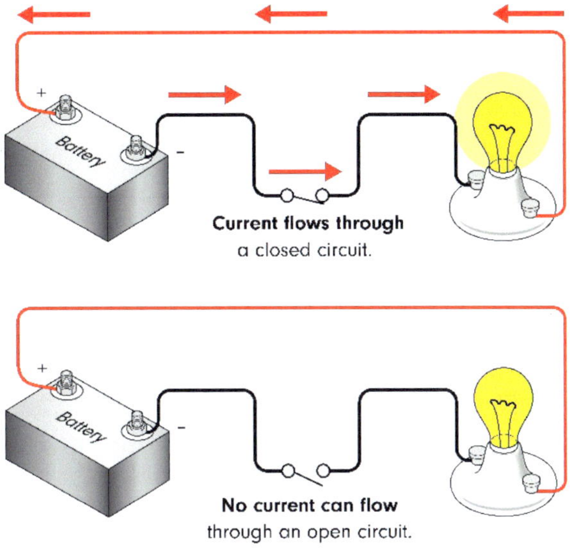

DC and AC Current

In circuits with cells or batteries the electrons are attracted to the positive terminal. The current is always moving in the same direction. This is called DC current.

In mains electricity the direction of current keeps changing direction. This is called alternating current.

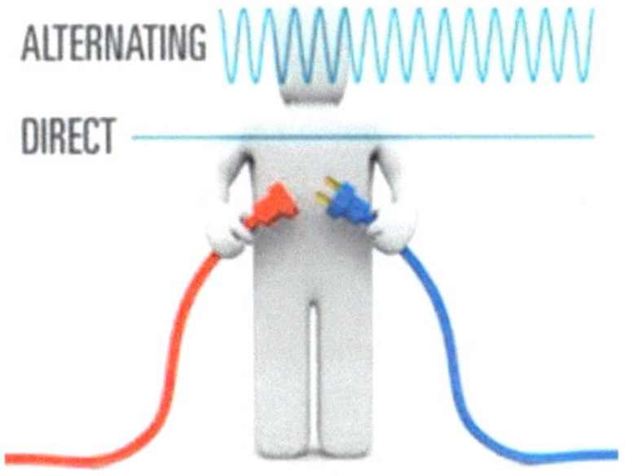

In the UK the mains electricity has a frequency of 50Hz and is 230 Volts. This means that it changes direction 50 times each second.

6.2 Domestic Electricity

6.2.1 Electricity in the home

At the end of this topic you should know:

- Most electrical appliances are connected to the mains using a three-core flex.
- The insulation covering each wire in the flex is colour-coded for easy identification:
    - live wire - brown
    - neutral wire - blue
    - earth wire - green and yellow stripes.
- The earth wire is a safety wire to stop the appliance becoming live and the fuse contains a thin piece of wire, which melts if the current becomes too large, thereby cutting off the supply. The resistance of a component is a measure of how difficult it is for an electric current to pass through it.
- Be able to select the correct fuse from a list when given the current rating of an appliance.
- Some appliances do not have an earth wire because they are double insulated.
- Everyday electrical appliances are designed to bring about energy transfers.
- The amount of energy an appliance transfers depends on how long the appliance is switched on for and the power of the appliance.
- Be able to use the equation:
  energy (kWh) = power (kW) x time (h)

Three Core Flex

Three core flex is the name given to standard electrical cable. It is called:

- Three core because it contains three wires
- Flex because it is flexible.

It is with three core flex that appliances are connected to the mains electricity.

Electrical appliances such, as these, are connected to the mains with three core flex.

Each wire is insulated with plastic to stop the current from moving from one wire to another. The insulation on these wires are colour coded:

- Brown – live wire
- Blue – neutral wire
- Green and yellow stripes – earth wire.

Each of these three wires have a different function.

The Earth wire is a safety feature. The wire goes into the earth, taking electricity away from the appliance. This prevents electric shock from the appliance.

This picture shows the earth wire taking the electricity away, to the ground.

The electrical plug at the end of the tree core flex, also contains a fuse.

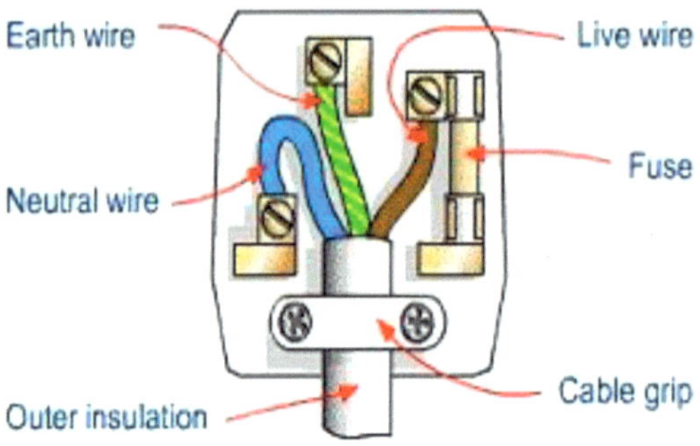

The fuse contains a very thin piece of wire that melts if too much current passes through it. When the fuse melts is breaks the circuit, stopping the current. This turns off the electricity and the appliance will no longer work.

Different fuses allow different currents through them before they melt. The fuse that is used needs to be higher than the maximum current of the appliance but no larger than necessary.

So, when choosing a fuse, choose the smallest fuse that allows the current to flow, when it is operating properly.

If the current used is:

- Less than 3 amps, use a 3amps fuse
- More than 3 amps but less than 5 amps, use a 5 amp fuse.
- More than 5 amps use a 13 amp fuse.

Some appliances do not have an earth wire. These appliances are double insulated. The double insulation prevent the electrical current from reaching the outside of the appliance and therefore no electrical shock can occur.

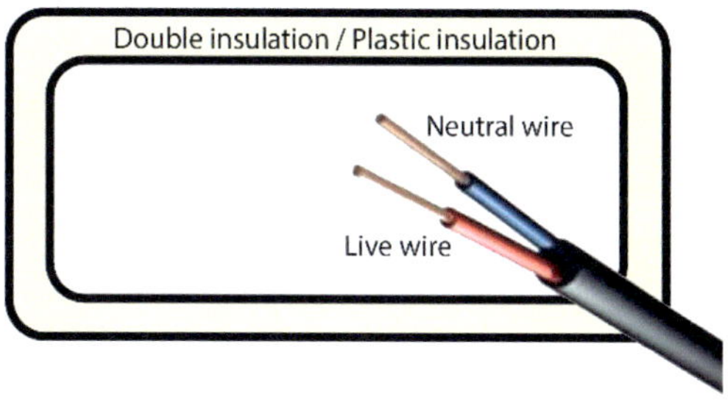

Energy Transfers

Appliances transfer energy. They use energy and they get that energy from the electricity.

For example, a kettle, heater or iron transfers electrical energy to heat energy, while a light bulb converts electrical energy to light energy.

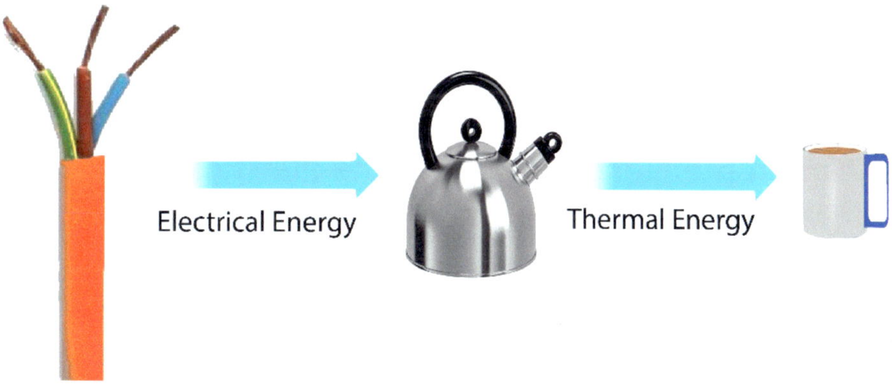

The amount of energy an appliance uses depends on two things:

- The power of the appliance
- The time it is on for.

Power is measured in watts (W) and is the amount of energy used per second.

1 Watt = 1 Joule of energy used / second

Therefore, we can work out the power using this equation:

$$power = \frac{energy}{time}$$

This can also be written as:

$$energy = power \times time$$

Sometimes energy is measured in kilowatt hours (kWh) and power in kilowatts (kW). A kilowatt is 1000 Watts. Notice that the W in the symbols is capital because it is named after a person. We can then write the equation is:

$$energy\ (kWh) = power(kW) \times time(h)$$

When answering a question that uses an equation there are four steps.

1. Write down the equation.
2. Substitute in the numbers given in the question.
3. Calculate the answer. Remember, you can use a calculator.
4. Write the answer, remembering to include the units.

For example, a 4kW clothes dryer is used for half an hour. How much energy is transferred?

Energy = power x time

= 4 x 0.5

= 2 kWh

6.3 Magnetism and Electromagnetism

6.3.1 Magnetism

At the end of this topic you should know:

- The poles of a magnet are the places where the magnetic forces are strongest.
- When two magnets are brought close together, they exert a force on each other.
- Two like poles repel each other. Two unlike poles attract each other.
- Attraction and repulsion between two magnetic poles are examples of non-contact force.
- The patterns of magnetic fields between bar magnets.

Magnetism

Magnetism is a type of non-contact force. That means that they can exert a force at a distance.

Magnets attract magnetic materials. Iron, nickel and cobalt are the only magnetic elements. Steel is magnetic because steel is mostly iron.

Magnets

Rectangular magnets are caller bar magnets.

The ends of a magnet are called poles. There is a North Pole and a South Pole.

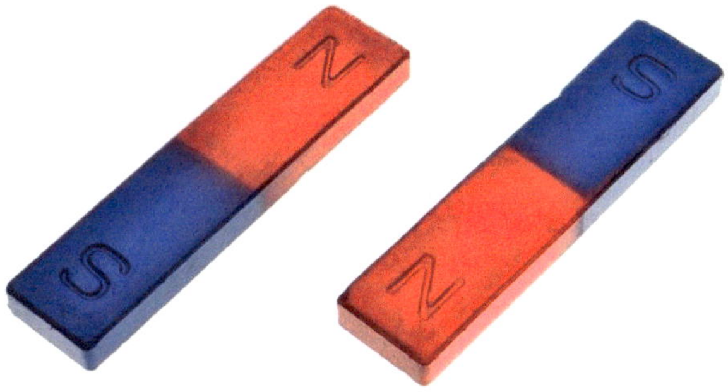

The poles of the magnet is where the magnetic forces are the strongest.

The law of magnetism states that:

- Like poles repel
- Unlike poles attract

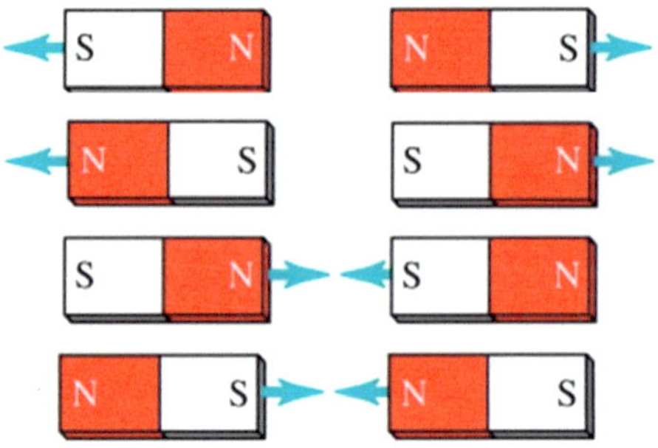

Magnetic Fields

The area around a magnet where there is a magnetic force is called the magnetic field.

The magnetic field can be seen by using very small pieces of iron, called iron filings. These arrange themselves in magnetic field lines.

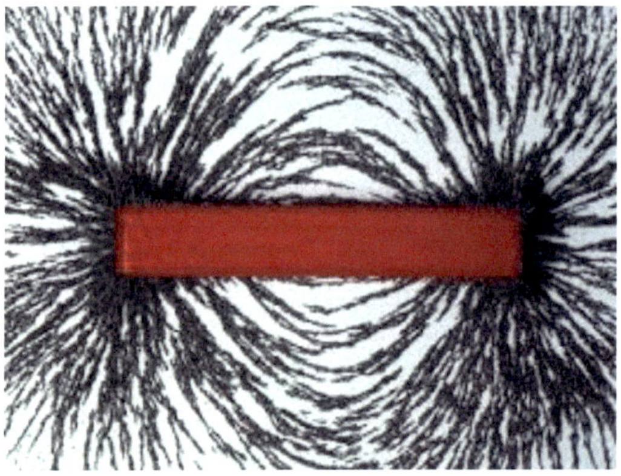

When we draw magnetic field lines, we always draw them with a single arrow on each line, that points towards the South pole.

Magnetic Field of a Bar Magnet

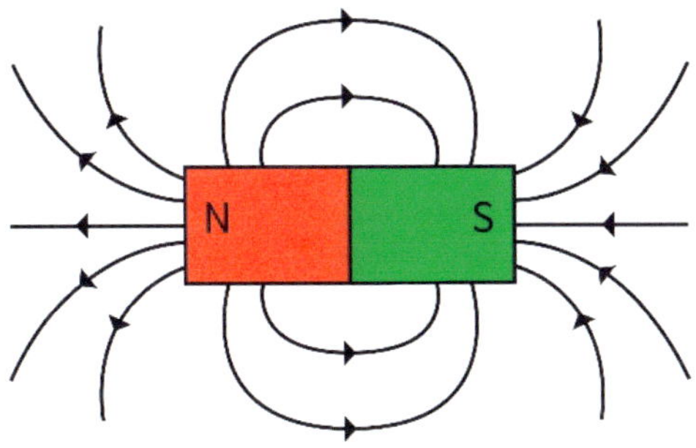

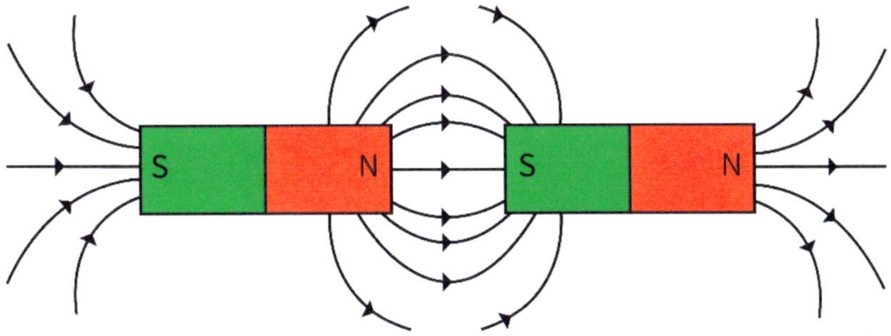

Attraction between opposite poles

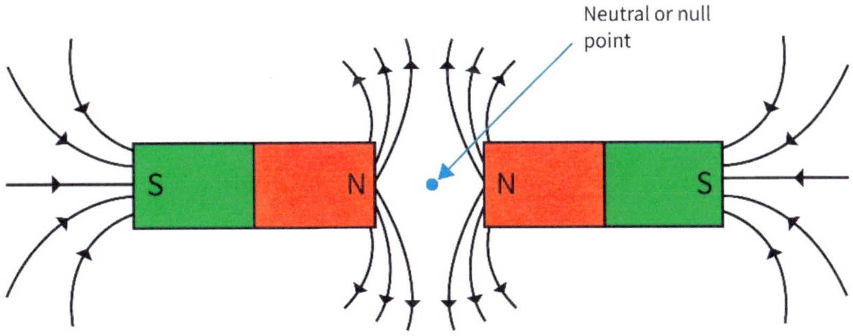

Repulsion between like poles

6.3.2 Electromagnetism

At the end of this topic you should know:

- When a current flows through a conducting wire a magnetic field is produced around the wire.
- The strength of the magnetic field depends on the current through the wire and the distance from the wire.
- Shaping a wire to form a solenoid increases the strength of the magnetic field created by a current through the wire.
- Adding an iron core increases the magnetic field strength of a solenoid.
- An electromagnet is a solenoid with an iron core.
- Know some common uses of electromagnets, e.g. in scrapyard cranes and relays.

When a current flows through a wire, a weak magnetic field is formed around that wire.

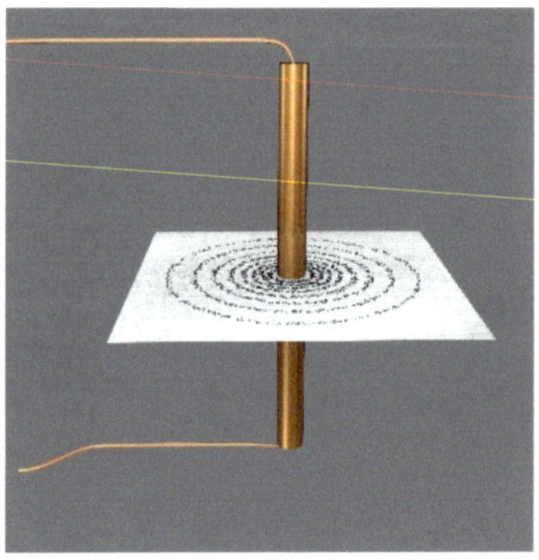

The strength of the magnetic field depends on the current flowing through the wire and the distance from the wire.

A coil of wire is called a solenoid.

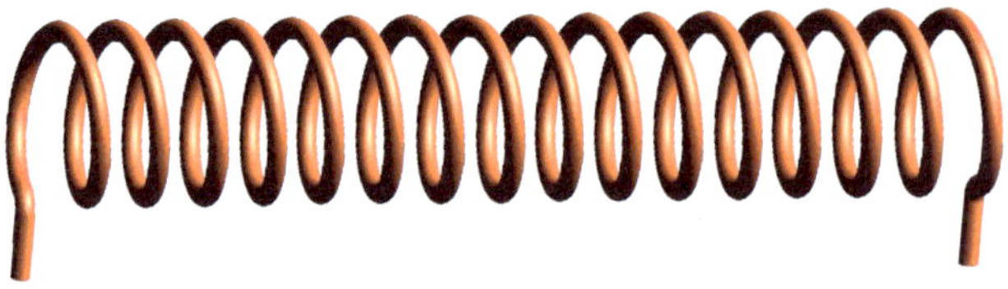

The magnetic field around a coil of wire is much stronger than through a wire. The magnetic field around a solenoid is the same shape as the magnetic field around a bar magnet.

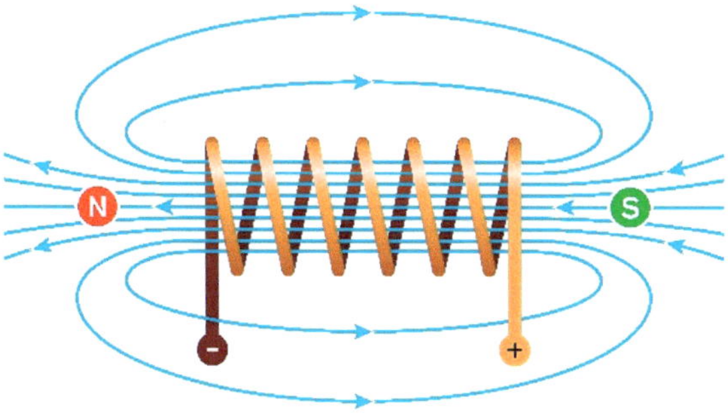

The strength of an electromagnet can be increased by:

- More coils
- Adding a soft iron core
- Higher current.

Electromagnets have lots of useful, because the magnetism can be turned on and off by turning on or off the current.

Scrapyards use electromagnets to pick up iron or steel, as when the electricity is turned off the iron or steel is released. This is useful for things like cars but not so useful for plastic or aluminium, as it will only work with magnetic materials.

Relay switches are a type of switch that uses electromagnets. They are very useful as a low voltage, low current circuit can be used to turn on a high voltage, high current circuit. This means that the person is not in contact with the circuit with the high current.

6.4 Different Types of Waves

6.4.1 Transverse and Longitudinal

At the end of this topic you should know:

- Waves may be either transverse or longitudinal.
- In a transverse wave the oscillations are perpendicular to the direction of energy transfer. The ripples on a water surface are an example of a transverse wave.
- In a longitudinal wave the oscillations are parallel to the direction of energy transfer. Longitudinal waves show areas of compression and rarefaction. Sound waves travelling through air are longitudinal.

Waves

Waves transfer energy from one place to another *without* the transfer of matter. Waves transfer energy with oscillations (the moving back and forth) of particles.

There are two types of waves:

- Transverse – the particles move at right angles to the direction of the wave. That is the oscillations are perpendicular to the direction of energy transfer.
- Longitudinal – the particles travel in the same direction as the wave. That is the oscillations are parallel to the direction of energy transfer.

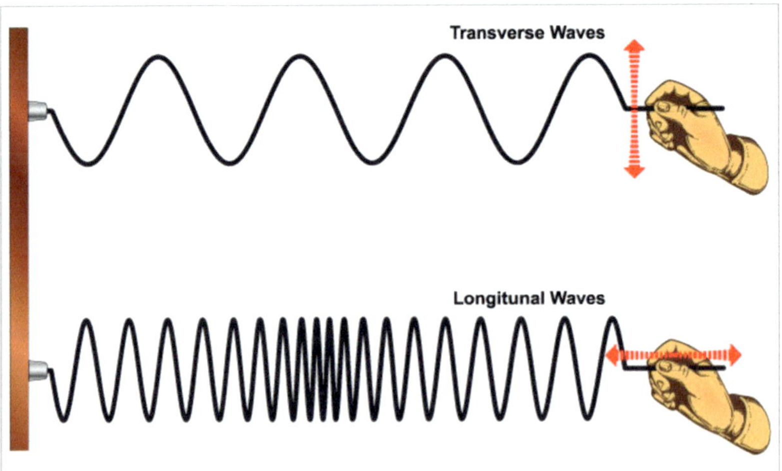

Examples of transverse waves include light and the ripples on the surface of water. If a buoy is on the water, the buoy will go up and down, but not move in the direction that the wave is moving.

Sound is an example of a longitudinal wave. In longitudinal waves the particles come close together, called a compression and further apart called a rarefaction.

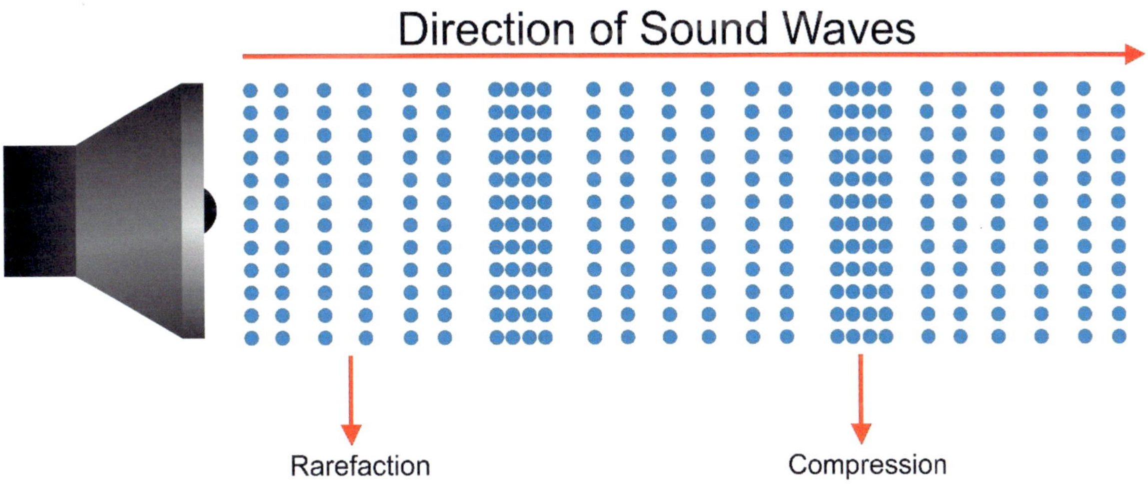

6.4.2 Describing Waves

At the end of this topic you should know:

- Waves are described by their amplitude, wavelength and frequency.
- The amplitude of a wave is the maximum displacement of a point on a wave away from its undisturbed position.
- The wavelength of a wave is the distance from a point on one wave to the equivalent point on the adjacent wave.
- Use the equation: wave speed (m/s) = frequency (Hz) x wavelength (m)
- The frequency of a wave is the number of waves passing a point each second.

Waves can be described by their amplitude, wavelength and frequency.

- Amplitude is the distance a particle moves from the middle of the wave. In transverse waves the amplitude is ***half*** the height of the wave. Amplitude is measured in metres.
- Wavelength is the distance between a point on a wave and the same point on the next wave. Wavelength is measured in metres.
- Frequency is the number of waves that pass a particular point in one second. Frequency is measured in Hertz.

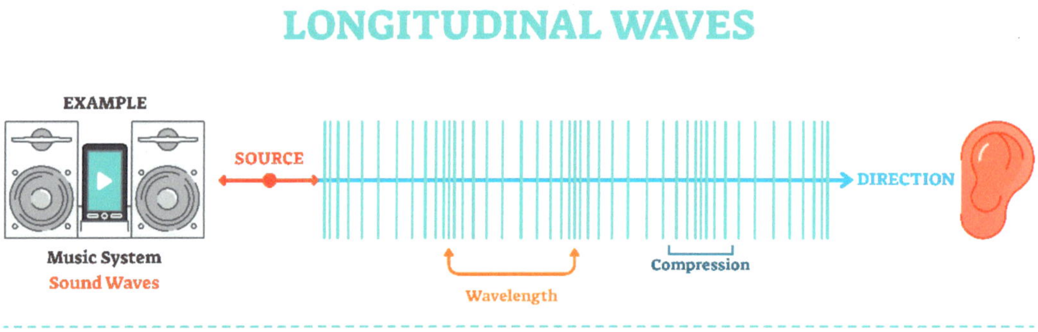

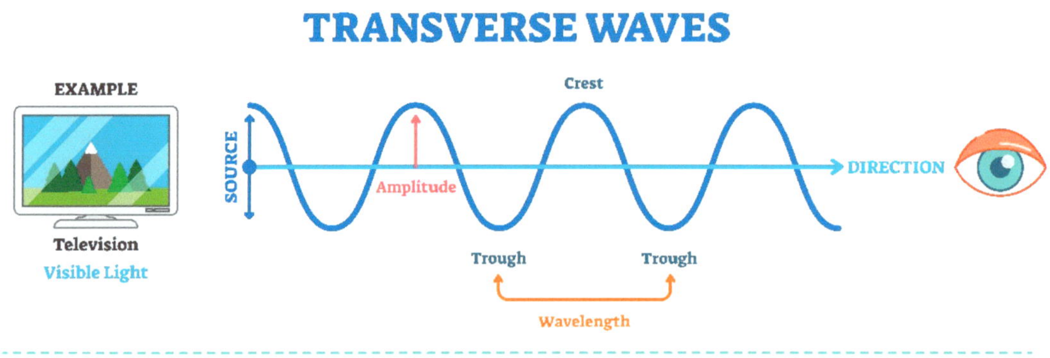

To calculate wave speed in metres per second (m/s), the following equation can be used:

$$wave\ speed = frequency \times wavelength$$

To work out the wavelength, there are four steps.

1. Write down the equation.
2. Put in the numbers that you are given in the question.
3. Calculate the answer. Remember, you can use a calculator.
4. Write down the answer. Remember to include the units.

For example, to work out what the wave speed is for a wave that has a frequency of 50 Hz and a wavelength of 0.2m.

wave speed = frequency x wavelength

= 50 x 0.2

= 10 m/s

6.5 Electromagnetic Waves

6.5.1 Electromagnetism

At the end of this topic you should know:

- Electromagnetic waves are transverse waves that transfer energy from the source of the waves to an absorber.
- Electromagnetic waves form a continuous spectrum and all types of electromagnetic wave travel at the same velocity through a vacuum (space) or air.
- The waves that form the electromagnetic spectrum are grouped in terms of their wavelength and their frequency.
- Going from long to short wavelength (or from low to high frequency) the groups are:
    - radio
    - microwave
    - infrared
    - visible light (red to violet)
    - ultraviolet
    - X-rays
    - gamma rays.
- Ultraviolet waves, X-rays and gamma rays can have hazardous effects on human body tissue. The effects depend on the type of radiation and the size of the dose.
- Electromagnetic waves have many practical applications, e.g.:
    - radio waves - television and radio (including Bluetooth)
    - microwaves - satellite communications, cooking food
    - infrared - electrical heaters, cooking food, infrared cameras
    - visible light - fibre optic communications
    - ultraviolet - energy efficient lamps, sun tanning
    - X-rays - medical imaging and treatments
    - gamma rays - for sterilising.
- Give brief explanations of why each type of electromagnetic wave is suitable for the practical application.

Electromagnetic Waves

Electromagnetic waves are transverse waves that transfers energy to a person, living organism or object that absorbs the energy. That is, they go from a source to an absorber. For example, some vegetables are placed in a microwave. Energy is transferred to the food by microwaves, a type of electromagnetic wave. The energy heats up the vegetables.

There are different types of electromagnetic waves. The different types of electromagnetic waves have different wavelengths and frequency. The velocity of all types of electromagnetic waves is the same in a vacuum or in air.

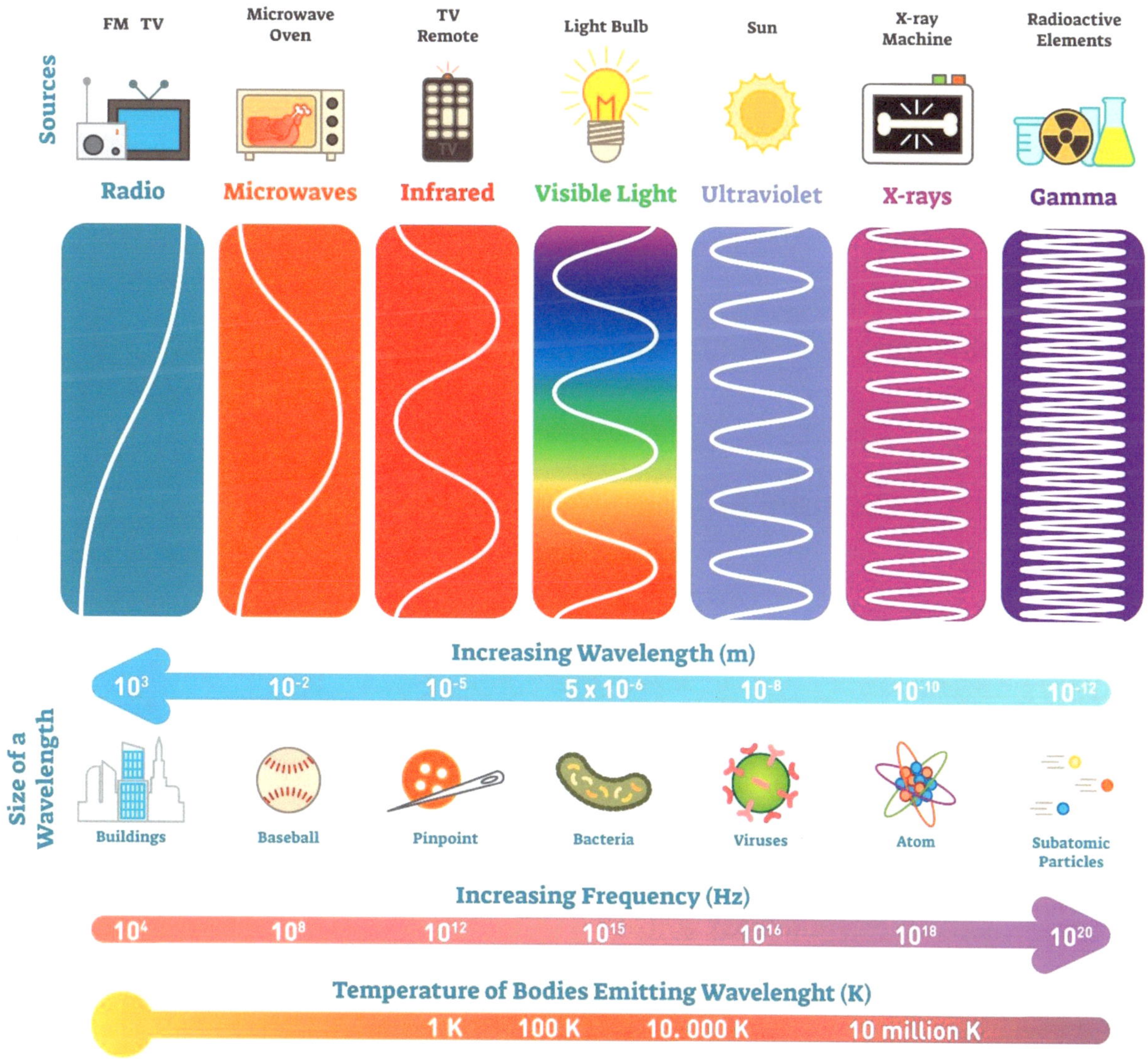

These types of wave form a continuous spectrum called the electromagnetic spectrum.

Going from longest wavelength and shortest frequency to shortest wavelength to longest frequency.

Radio

Radio waves have the longest wavelength. They are used to transmit radio signals and TV signals. Bluetooth also uses radio waves.

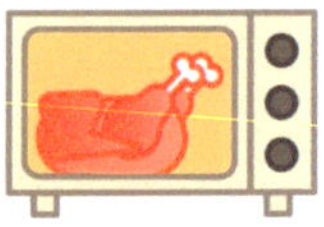

Microwaves

The most obvious use of microwaves is the heating of food and drink. Microwaves are more energy efficient than other cooking equipment and they heat food faster.

Microwaves are also used for satellite communication.

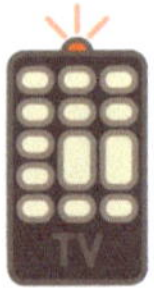

Infrared

Infrared is used in electrical heaters, cooking equipment, infrared cameras used to detect heat and in TV remotes.

Visible Light

Visible light is the only part of the electromagnetic spectrum that we can see.

Visible light is used in cameras and in fibre optic communications.

Ultraviolet

Ultraviolet light, also called UV, is emitted by the sun and is used by people to get a suntan. However, too much UV can damage eyes and cause skin cancer.

UV light is also used in security markers.

X-rays

X-rays are used in medical imaging. X-rays can damage human tissue, so those that work with x-rays need to monitor how much radiation they receive. This is because the amount of damage depends on the dose received. X-rays are also dangerous for unborn babies.

Gamma

Gamma radiation has the smallest wavelength and the highest frequency of the different types of electromagnetic waves.

Gamma radiation is used to sterilise medical equipment and in cancer treatment.

Gamma radiation can kill cells. This is why they work in sterilisation is it kills the microbes and in cancer treatment as it kills the cancer cells. However, care must be taken to minimise the dose of gamma radiation received and those who work with gamma radiation must monitor how much radiation they receive.

Ultraviolet, X-rays and gamma waves can have hazardous effects on human body tissue. The effects depend on the type of radiation and the size of the dose.

Credits

Page 80. Images of iron and sulfur by Asoult and is available on Creative Commons at: https://commons.wikimedia.org/wiki/File:Fe-S_mixture_05.jpg and https://commons.wikimedia.org/wiki/File:Fe-S_mixture_02.jpg

Page 82. Image of salt water solution by Chris 73 and is available on Creative Commons at: https://commons.wikimedia.org/wiki/File:SaltInWaterSolutionLiquid.jpg

Page 84. Image of distillation equipment is adapted from the image by Quantockgoblin which is available from Creative commons at: https://commons.wikimedia.org/wiki/File:Simple_distillation_apparatus.svg

Page 159. Image of solenoid is by Zureks who released it into the public domain. It is available at: https://commons.wikimedia.org/wiki/File:Solenoid-1.png

Page 149. Image illustrating the roles of voltage, current and resistance. Drawn by Tahir626.

Made in the USA
Las Vegas, NV
03 November 2026

57422813R00102